Agricultural Growth for Poverty Alleviation

N.K. Sinha

RANDOM
RANDOM PUBLICATIONS
NEW DELHI (INDIA)

Agricultural Growth for Poverty Alleviation

ISBN 978-93-5111-112-2

Published in 2013 in India by

RANDOM PUBLICATIONS

4376-A/4B, Gali Murari Lal, Ansari Road
New Delhi-110 002
Phone : +9111-43580356, 011-23289044
e-mail: randompublications@hotmail.com

Reprinted 2023

Type Setting by : Friends Media, Delhi-110089
Digitally Printed at : Replika Press Pvt. Ltd.

Preface

The major sources of agricultural growth during this period were the spread of modern crop varieties, intensification of input use and investments leading to expansion in the irrigated area.

Poverty is a rural phenomenon in most of the developing world, especially the low-income developing countries. The rural poor make up more than 75 percent of the poor in many Sub-Saharan African and Asian countries. Accelerated public investments are needed to facilitate agricultural and rural growth. These investments need to be supported by an enabling policy environment. This includes trade, macro, and sectoral policies that do not discriminate against agriculture, and policies that provide appropriate incentives for the sustainable management of natural resources.

To date, rural poverty alleviation remains a priority issue in most countries of the Asian and Pacific region. Important in this respect is the enhancement of rural employment and income generation. The recent financial crisis in Asia has increased the burden on rural areas to (re)absorb labourers who lost employment in urban centres. A potential strategy to increase rural employment as well as enhance livelihood options for rural families is the development of opportunities in organic agriculture. Organic agriculture has shown to help improve soil conditions and farm health for small farms. It may provide for a diversified income base through diversification of crops, and uses more labour compared to conventional agricultural methods *i.e.* for planting, strip-farming, non-chemical weeding and composting. Besides, processing and marketing of organic agricultural products could create new off-farm rural employment opportunities, especially when lucrative export markets can be accessed.

This book brings the comprehensive coverage of the responsible factors for the persistence of poverty in the country as well as highlights some crucial issues to be addressed to tackle the menace of poverty.

—*Author*

Contents

1

Introduction

Poverty has fallen rapidly over the past 40 years, but at different rates around the world. Asia has achieved the most rapid poverty reduction, particularly China, but also India and South East Asia. In contrast, little if any progress has been made in sub-Saharan Africa, where the number of people living on less than one dollar a day – the internationally agreed definition of absolute poverty – has doubled over the past 20 years. Historically, rates of poverty reduction have been very closely related to agricultural performance – particularly to the rate of growth of agricultural productivity. In simple terms, this indicates that the countries that have increased their agricultural productivity the most have also achieved the greatest reductions in poverty.

In Asia rapid productivity gains, achieved largely through the technological advances of the Green Revolution, provided a 'fast-track' route out of poverty by directly increasing producers' incomes and labourers' wages, by lowering the price of food and by generating new livelihood opportunities as success in agriculture provided the basis for economic diversification. However, despite decades of investment in new agricultural technology and rural development, hunger and poverty continue to plague large areas of the developing world. The problem is particularly acute where people depend on rain-fed agriculture, in particular sub-Saharan Africa, where the impact of new technologies has been less apparent and agricultural productivity has generally stagnated and even fallen in some areas.

Achieving the Millennium Development Goal (MDG) of halving the proportion of people living in absolute poverty by 2015 will depend largely on increasing agricultural productivity, which remains perhaps the single most important determinant of economic growth and poverty reduction. This fact is not lost on developing countries or their development agency partners, who are seeking ways to stimulate agricultural development. But serious doubts are emerging as to whether agricultural productivity can be increased where it is needed most, and what part, if any, small-scale farming will play in the future. Underlying such doubts is a concern that the context in which small-scale agriculture could achieve productivity gains today is very different to that which prevailed in Asia during the halcyon days of the Green Revolution. Inherent differences in production capabilities and quite fundamental changes in the international agricultural context are combining to create a set of circumstances that are probably far less conducive to achieving the type of productivity transformation witnessed in Asia.

SUSTAINABLE AGRICULTURE AND POVERTY ALLEVIATION

To date, rural poverty alleviation remains a priority issue in most countries of the Asian and Pacific region. Important in this respect is the enhancement of rural employment and income generation. The recent financial crisis in Asia has increased the burden on rural areas to (re)absorb labourers who lost employment in urban centres. A potential strategy to increase rural employment as well as enhance livelihood options for rural families is the development of opportunities in organic agriculture. Organic agriculture has shown to help improve soil conditions and farm health for small farms. It may provide for a diversified income base through diversification of crops, and uses more labour compared to conventional agricultural methods *i.e.* for planting, strip-farming, non-chemical weeding and composting. Besides, processing and marketing of organic agricultural products could create new off-farm rural employment opportunities, especially when lucrative export markets can be accessed.

Until the early 1990s, organic agriculture remained a relatively rare phenomenon with less than 1% of farmers involved in most countries. Production, marketing and consumption of organic agricultural produce depended heavily on private initiatives. Over the last decade, organic agriculture has seen a considerable boost particularly in developed countries. Also in many developing countries in Asia, small and medium scale initiatives have been undertaken to advance organic farming. However, government support has remained limited.

National governments can play an important role in encouraging the adoption of organic agriculture in developing countries, particularly in the following areas: research and development, education and extension into organic agriculture methods, legalization and certification for organic produce, pricing policies for agricultural input and output, land-tenure, and export promotion.

AGRICULTURE GROWTH RATE

Agriculture Growth Rate in India GDP had been growing earlier but in the last few years it is constantly declining. Still, the Growth Rate of Agriculture in India GDP in the share of the country's GDP remains the biggest economic sector in the country. In the 1980s, Indian policymakers shifted their focus from food self sufficiency to generating additional income in rural areas as a means of tackling the problem of poverty, which was concentrated in rural areas. Acceleration of agricultural growth, with a special focus on improving the position of small farmers and extending the productivity revolution to non-irrigated areas was seen as a critical part of the strategy for poverty alleviation. This effort was supplemented targeted anti poverty programmes to address the needs of vulnerable groups who may not benefit sufficiently from general agricultural growth. India achieved considerable success with this approach in the 1980s. Growth of agricultural gross domestic product (GDP) accelerated to about 4.7 percent in the 1980s, compared with only 1.4 percent in the 1970s. This, agricultural growth, together with the beginning of economic reforms in the nonagricultural sector, pushed up the growth rate of overall GDP to around 5.8 percent in the period 1980-81 to 1989-90 compared with about 3 percent in the 1970s.

India's growth was disrupted at the start of the 1990s by a major balance of payments crisis which led to the adoption of an extensive process of structural reforms. It took time to regain momentum and it was only in 1993-94 that the economy got back on track, clocking an average growth rate of 6.8 percent in the three years 1993-94 to 1995-96. This acceleration in growth in the post reform period led policymakers to set a more ambitious GDP growth target of 8 percent a year for the Ninth Plan period (1997-98 to 2001-2002), to be supported by a growth rate of 4 percent a year in agriculture. The projected growth of 4 percent per year in agriculture was clearly inline with the average growth of 3.8 percent achieved in the period 1990-91 to 1996-97.

However, actual performance since the mid 1990s has been disappointing. Agricultural growth slowed to 2 percent a year in the Ninth Plan period, and overall economic growth was only 5.5 percent, well below the 8 percent target. Since agriculture accounted for about 25 percent of GDP, the shortfall of more than 2 percentage points in agricultural GDP growth compared with the target accounts directly for a shortfall of about half a percentage point in GDP growth. If the indirect effects of more rapid agricultural growth on other sectors are taken into account, the total impact on GDP growth may have been as much as one percentage point.

These shortfalls were known, when the Tenth Plan (covering the period 2002-03 to 2006-07) was formulated, but it was assumed that the poor performance of agriculture was due to temporary factors such as poor monsoons and depressed agricultural commodity prices in world markets following the East Asian meltdown. The Tenth Plan therefore adopted the same targets of 8 percent growth in GDP and 4 percent growth in agriculture. Experience in the first three years of the Tenth Plan period has sounded some alarm bells. GDP growth has averaged about 6.5 percent, but agricultural GDP in these years (2002-03 to 2004-05) has grown by only 1.1 percent per year.

The loss of dynamism in agriculture explains most of the shortfall in aggregate GDP growth.

Slower growth in agriculture also has direct implications for poverty reduction in rural areas. Official figures suggest that the incidence of poverty fell from 36 percent in 1993-94 to 26 percent in 1999-2000. The comparability of these numbers has been questioned because of recent changes (ostensibly improvements) in the methods for measuring consumption in household surveys, but there is a broad consensus that if corrections are made to ensure comparability, the percentage of the population in poverty has declined significantly, though less than in the official figures. However, even the official figures show less decline than what had been targeted, and this is undoubtedly a reflection of the slowdown in agricultural growth. Slow growth in agriculture is also at the root of growing evidence of distress in the farming community. Surveys show that a large percentage of farmers want to leave farming because they find it is no longer sufficiently profitable. The uncertainty associated with farming has also increased and farmers lack effective means of insuring against such risks. There are larger market uncertainties associated with new crops and poultry because of greater vulnerability because of falling ground water levels. There is evidence of increased indebtedness arising from the inability to cope with risks.

Recognizing these problems, the Government has undertaken a comprehensive review of the strategy for agriculture in order to come up with a new deal for agriculture and the rural economy in general. Remedial action will be needed on several fronts including increased public investment in irrigation and rural roads, better management of existing irrigation systems and of water resources in dry land areas, a strengthened agricultural research system and more effective extension, improvements in the production and distribution of certified seeds, improvements in the credit delivery system, and innovative steps in marketing and contract farming to support the diversification of Indian agriculture.

IRRIGATION

Water is a critical constraint on raising agricultural productivity and much of the success of the Green Revolution came from improved productivity in areas of assured irrigation provided through canals or (much more significant) through ground water utilization. The scope for expanding irrigation through large and medium scale projects has yet to be fully exploited. Out of the total of 59 million hectares that could be irrigated through such projects only 40 million hectares have been irrigated. The slow pace of exploitation of irrigation potential is due to lack of resources in state governments and the tendency to spread available resources thinly over too many projects. Additional public investments in this area are therefore essential for early utilization of the potential.

Effective maintenance of the existing system of canal irrigation also suffers because the irrigation departments of the states lack resources. This in turn is because water charges are kept too low, covering only 20-25% of the operations and maintenance cost of the system in most states. Poor maintenance leads to loss of water through seepage, with the result that water use efficiency is very low – around 25 to 40 percent instead of 65 percent that should be attainable. Low water charges also encourage highly water intensive crops at the upper end of the canal network, leaving tail-end portions starved of water.

The solution lies in rationalization of water rates to ensure adequate financial resources to cover maintenance and resort to participatory irrigation management to give farmers a stake in the operation and maintenance of the system. Some interesting experiments in these have promise. Maharasthra has recently established a Water Regulatory Authority to set water charges in a non-political manner. Several states are also experimenting with involving water user associations (WUAs) in the operation of the canal systems. Ideally the WUAs should be empowered to collect water charges and to retain a portion of the collection to maintain the portion of the distribution network operating in their area.

Ground water utilization played a major role in expanding irrigation in the 1980s but uncontrolled exploitation of groundwater has led to serious depletion of the water table in many parts of the country. Overexploitation is encouraged by the policy of massive under pricing of electricity for agricultural use, with a few states having made electricity for farmers completely free. Even where it is not free, the charge for electricity is a fraction of the average cost, and is not based on metered use. Instead there is a fixed charge for presumed usage, based on the capacity of the pump, an arrangement which implies that the marginal cost of electricity for pumping ground water is zero. Under pricing canal water and electricity are clearly highly distortionary, given the need to conserve water use. They are also distributionally unfair because the benefits of under priced water accrue disproportionately to upper end farmers whereas under priced power enables those able to afford larger pumps to lower the water table denying water to farmers who can only afford shallow wells.

The investment requirements of irrigation are massive. Completion of all unfinished projects alone is estimated to cost approximately US$ 20 billion. In addition, provision must be made for new irrigation projects (large, medium and small), which together will require about US$ 45 billion. The total requirement is therefore about US$ 65 billion.

WATER MANAGEMENT IN RAINFED AREAS

Water management is the activity of planning, developing, distributing and managing the optimum use of water resources. In an ideal world, water management planning has regard to all the competing demands for water and seeks to allocate water on an equitable basis to satisfy all uses and demands. This is rarely possible in practice. About 60 percent of India's cultivable area will remain dependent on dry land farming even after all irrigation potential is fully exploited. Productivity growth in these areas is obviously critical for rural income growth and poverty alleviation, and it depends critically upon better moisture conservation and the development of varieties suited to dealing with moisture stress.

Schemes for water retention, moisture conservation and groundwater recharge have been implemented for many years in India but with mixed results. Experience suggests some pointers for the future. Greater use of technology inputs can help a great deal. Satellite mapping by the Indian Space Research Organisation (ISRO) has been particularly helpful in planning watershed management schemes to achieve optimal results. It is also important to adopt a holistic approach. For example, if deforestation problems upstream are not tackled, water retention structures downstream will quickly silt up. Community participation is critical to impart ownership and ensure an acceptable distributional outcome. In the past, these multiple factors were not effectively integrated into watershed development schemes. Now a National Rainfed Area Authority has been proposed to help coordinate the work of different implementing agencies.

The cost of treating rainfed areas to ensure optimum use of available water is approximately Rs.10,000 per hectare and the untreated area is about 80 million hectares, yielding a total cost of approximately US$20 billion. If this amount is added to the cost of irrigation development and the target to be achieved over a 10-year period it would require a doubling of public investment in irrigation.

OTHER INPUTS

Increasing agricultural productivity also depends on the efficient delivery of several other inputs. The quality of seeds and planting material needs to be greatly improved, and this calls for strengthening the research effort to make it more effective. Two expert committees have recently reported on how to restructure the agricultural research system to make it more results oriented and their recommendations are under consideration. The system for producing and marketing certified seeds of existing varieties at reasonable prices also needs to be improved. Seed replacement rates in most parts of the country are only one-third to one-half of what they should be, a situation which reflects partly a lack of knowledge of the importance of seed replacement and partly also a lack of availability of reliable seeds.

There is evidence that the use of fertilizers is at present highly imbalanced suggesting that scientific application of fertilizers holds potential for raising productivity. Nitrogen fertilizers are oversubsidized compared with phosphorus and potassium fertilizer. The structure of fertilizer subsidies should be rationalized to avoid excessive and wasteful use of nitrogen fertilizers. Lack of knowledge of micronutrient deficiency in the soil is also a serious problem.

There is need for much more extensive soil testing to encourage balanced application of nutrients needs. Underlying these problems is the deterioration of the extension services, which makes it difficult to disseminate best farming practices. Strengthening the extension system therefore needs special attention. The government has also identified credit to farmers as a critical area for corrective action. The public sector commercial banks are being pushed to provide credit to agriculture and have responded commendably. However, the cooperative credit system which was meant to be the backbone of agricultural credit, has become financially weak. Part of the problem has been the politicization of cooperative institutions as a consequence of interference by state governments. The central government is considering ways of reviving the cooperative credit system by recapitalizing the cooperative banks provided state governments agree to changes in the system of governance that would ensure professional management of cooperative banks without state government interference.

AGRICULTURAL DIVERSIFICATION

India's future agricultural strategy must also be oriented to the need for agricultural diversification. India's foodgrain production capacity has increased significantly over the years and there is evidence that household consumption patterns are changing away from foodgrain towards higher value crops such as vegetables, fruits, milk, eggs, etc. Future growth in agriculture must come from diversification into these non-food grain areas and this will pose a special challenge because marketing these perishable products is much more complicated than marketing food grains.

Horticulture development is currently constrained by poor marketing arrangements. The gap between prices received by the farmers and those paid by urban consumers is large, reflecting inefficient marketing arrangements. Horticultural produce is typically collected from farmers by market agents, who sell it organised markets established under the Agricultural Produce Marketing Acts. Unfortunately these markets are controlled by a few traders and operate on highly non transparent basis. Facilities for grading and handling are poor and methods of price discovery in the markets are nontransparent. Wastage is high owing to poor logistics and the absence of cold chains. The net result is much lower realisation by the farmer. It is necessary to amend outdated laws restricting the establishment of markets to allow cooperatives and private entrepreneurs to set up modern markets with grading facilities, cold storages and transparent auction procedures. Half a dozen states have already amended their existing laws on agricultural marketing to allow such markets to be established and a dozen others are in the process of doing so. These changes are being resisted by those who control the existing structure but this opposition will weaken over time.

Contract farming is another innovation that has been introduced in many states and could accelerate diversification. India's laws on agricultural land do not allow corporate bodies to purchase land and operate large-scale farms – a national policy to prevent displacement of a large number of small farmers. – but corporate buyers who know what is needed in export markets, in high end domestic markets or in agro-processing, can engage in contract farming to procure high quality produce. Buyers select areas suitable for the crops they are interested in and organize farmers to produce these crops under contract, while providing planting material of the right quality and technical supervisory. The process enables the farmer to eliminate marketing risk while allowing the corporate buyer to ensure quality supplies by selecting planting material, and provide access to scientific advice on disease and other types of stress.

The development of agro-processing will be a spur agricultural diversification and the government is paying special attention to this area. At present, the proportion of India's agricultural output that is processed is very small compared with that in most developing countries and the demand for processed food is bound to increase as incomes rise. There are several obstacles to the more rapid development of food processing. Taxation structures often discriminate against food processing because processed food is the first stage for application of indirect taxes and the absence of tax rebate on taxes paid on inputs means the effective tax on value added is very high. Another impediment is the reservation of certain categories of products for small scale production. The absence of a modern food processing law has meant that this sector is governed by multiple laws, making it difficult to operate effectively. An Integrated Food Processing Law has been introduced in Parliament to replace and the passage expected in the current year will make a qualitative difference to the operating environment.

AGRICULTURAL DIVERSIFICATION ON SMALL FARMS

Agricultural diversification is an important mechanism for economic growth. It depends, however, on there being opportunities for diversification and on farmers' responsiveness to those opportunities. Agricultural diversification can be facilitated by technological breaks-through, by changes in consumer demand or in government policy or in trade arrangements, and by development of irrigation, roads, and other infrastructures. Conversely, it can be impeded by risks in markets and prices and in crop-management practices, by degradation of natural resources, and by conflicting socio-economic requirements-perhaps for employment generation, or for self-sufficiency or foreign-exchange-earning capacity in particular crops or livestock or fishery or forest products.

PRESENT STATUS OF AGRICULTURAL BIOTECHNOLOGY IN ASIA

SELECTED COUNTRY EXPERIENCES

Several governments in Asia are committed to the use of modern biotechnology in agriculture. They have devoted significant human and financial resources to this policy over the past two decades. The PRC accords high priority to biotechnology to increase food production and improve product quality in an environmentally sustainable manner. The PRC has moved quickly to adopt new biotechnologies, particularly genetically modified crops. The country is rapidly increasing its expenditure on biotechnology R&D. Over 103 genes have been evaluated for improving traits in 47 plant species. New traits have been introduced and evaluated in field tests on rice, wheat, maize, cotton, tomato, pepper, potato, cucumber, papaya, and tobacco. A variety of traits have been targeted. They include resistance to diseases, pests, and herbicides, and quality improvement. Approximately 50 genetically modified varieties have been approved for environmental release, or small-scale field testing in the PRC. A few new genetically improved varieties have been approved for largescale commercial production. The most widespread are new pest-resistant varieties of cotton that are being widely cultivated by farmers. These were grown commercially by approximately 3 million farmers on approximately 500,000 ha in 2000. Several new products are in the pipeline for potential commercialization. India has allocated large public resources towards human resources development and infrastructure in biotechnology. In the early 1980s, the Government of India created a Department of Biotechnology to promote the use of new biotechnologies in industry, medicine, and agriculture. Current R&D efforts in India are directed towards increasing agricultural productivity, bioremediation in the environment, medical and industrial biotechnology, and bioinformatics. R&D priorities in agriculture include new regeneration techniques for the rapid multiplication of citrus, coffee, mangrove, vanilla, and cardamom. Cardamom yield has increased 40 percent through the use of tissue culture.

There is substantial private sector participation in biotechnology in India, for example, in the seed sector, the veterinary products sector, and bioinformatics linked with the booming information technology sector. Further, organizations like the M.S. Swaminathan Research Foundation have developed innovative approaches such as the ADB-supported BioVillages programme that is fostering the growth of new income-raising technologies. Emphasis is on developing small-scale bioindustries for women.

Indonesia has placed a high priority on biotechnology over the past 15 years. The Government has designated three National Biotechnology Centers to coordinate R&D in agriculture, medicine, and industrial microbiology. Applications of biotechnology to agriculture are primarily the responsibility of the Agency for Agricultural Research and Development (AARD). A National Committee on Biotechnology advises the minister in developing guidelines for government policy in the promotion of biotechnology. In recent years there has been an extensive training programme within Indonesia and abroad to upgrade skills of scientists involved in biotechnological research. In the 1980s, a major World Bank loan of over $100 million financed the creation of three inter-university centers for agricultural, medical, and industrial biotechnology. More recently, a current World Bank loan is financing facilities for agricultural biotechnology within AARD.

Crop improvement efforts using modern biotechnology started in Pakistan in 1985, when a training course was held on recombinant DNA. Work is now concentrated on chickpea, rice, and cotton. Field evaluation is hampered by lack of biosafety regulations. There is some private investment in R&D of agricultural biotechnology. The government controls testing, multiplication, distribution, and biosafety issues for genetically modified crops. Pakistan lacks firm policy and regulations regarding intellectual property rights (IPR) and patents involving biotechnology, and biosafety regulations for GMOs.

The Philippines began its modern biotechnology programmes in 1980 with the creation of the National Institutes of Molecular Biology and Biotechnology in Los Baños, with a focus on agricultural biotechnology. In 1997, the Agriculture Fisheries Modernization Act recognized biotechnology as a major strategy to increase agricultural productivity. The Act provided a budget for agricultural biotechnology of almost $20 million Present Status of Agricultural Biotechnology in Asia 21 annually for the next 7 years (4 percent of the total R&D budget). In 1998, the government funded these five high-level biotechnology research projects to develop

(i) new varieties of banana resistant to banana bunchy top virus and papaya resistant to ringspot virus,
(ii) delayed ripening papaya and mango,
(iii) insect-resistant maize,
(iv) marker-assisted breeding in coconut, and
(v) coconut oil with high lauric acid content.

Nongovernment organizations (NGOs) and other groups concerned about the safety of GMOs have been vocal in the Philippines. This is affecting field-testing and commercialization of transgenic crops. Products in the regulatory pipeline include new varieties of insect-resistant maize and insect-and disease-resistant rice. Thailand is focusing on the applications of biotechnology to traditional foods, fruits, and export commodities such as shrimp. R&D priorities are to increase production and reduce production cost on crops such as rice, cassava, sugarcane, rubber, durian, and orchids. An early success in

Thailand has been the development of new molecular diagnostics for the diagnosis and control of virus diseases in shrimp. These diseases cost the shrimp export industry over $500 million in lost production in 1996. The development and commercial use of the new diagnostics prevents the loss of an estimated 20–50 percent of annual production, a saving of at least $100 million per year. There are also active agricultural biotechnology programmes in Bangladesh, Malaysia, Nepal, Sri Lanka, Singapore, and Viet Nam.

REGIONAL AND INTERNATIONAL PROGRAMS

International Agricultural Research Centers

International R&D programmes using modern biotechnology are being conducted by the international agricultural research centers (IARCs), particularly the International Rice Research Institute (IRRI), International Maize and Wheat Improvement Centre (CIMMYT), International Crops Research Institute for the Semi-Arid Tropics (ICRISAT), International Livestock Research Institute (ILRI), and International Service for National Agricultural Research (ISNAR). The Centre for International Forestry Research also uses biotechnology in the characterization of forest diversity in its Asian programme. The International Centre for Living Aquatic Resources Management is using new technologies in the improvement of fisheries and aquaculture systems. ILRI is initiating a programme on Asian livestock improvement. The CGIAR centers invest approximately $30 million per year in modern biotechnology.

ISNAR Biotechnology Service

The ISNAR Biotechnology Service (IBS), with Japanese support, has been assisting selected Asian countries in developing human resources for managing biotechnology research programmes or institutions. IBS has developed specialized courses to enhance the capacity and competency of managers, focusing on strategy building, priority setting, managing biosafety and regulatory aspects, resource generation and deployment, product delivery, and information sharing as well as the establishment and management of linkages.

Acquisition of Agri-biotech Applications

The International Service for the Acquisition of Agri-biotech Applications (ISAAA) is brokering public-private sector partnerships to facilitate technology transfer. It has current regional projects to increase the productivity of maize, papaya, and sweetpotato. ISAAA is also establishing a new Asian knowledge centre for crop biotechnology, based in the Philippines, in partnership with CAB International and the Southeast Asian Regional Centre for Graduate Study and Research in Agriculture. Its purpose is to make available timely and balanced information on the risks and benefits of crop biotechnology to interested parties in Asia. In doing so, it aims to provide training and study tours highlighting experiences not only in research but also with biosafety and intellectual property issues. Present Status of Agricultural Biotechnology in Asia

ACTIVITIES OF FUNDING AGENCIES

The major external sources of assistance for agricultural biotechnology in the Asia/Pacific region are ADB, Australia, the Rockefeller Foundation, the United Nations, the United States Agency for International Development (USAID), and the World Bank.

Asian Development Bank

ADB has made several strategic and innovative investments in agricultural biotechnology over the past decade. These investments have been in the form of loans and technical assistance projects.

Agriculture and Science and Technology Loan Projects

Several governments have requested ADB financial support for human resource development, laboratory facilities, and equipment for agricultural biotechnology programmes. These programmes have been integrating new applications of biotechnology into their conventional agricultural R&D programmes.

The applications include the use of new molecular diagnostics for pests and diseases and MAS for crop and livestock breeding. Such components are being supported under ongoing ADB projects in the Philippines, Sri Lanka, and Thailand. ADB has provided other loans and grants in closely related areas such as in integrated pest management in cotton under the Cotton Development Project in Pakistan.

ADB Regional Technical Assistance Projects

ADB has provided regional technical assistance grants for the development of the three regional crop biotechnology networks over the past decade. The networks are the Asian Rice Biotechnology Network (ARBN), initiated in 1993, the Asian Maize Biotechnology Network (AMBIONET), initiated in 1998 and the recently established Asian Semi-Arid Crops Network. These networks have been influential in developing capacity to use new techniques in crop breeding for the major cereal crops (rice, maize) and the crops important in the semi-arid regions (sorghum, pigeonpea, and groundnut). The networks are managed by three IARCs (IRRI, CIMMYT, and ICRISAT) that work with national re search institutes to implement the programmes on rice, maize, and semi-arid crops.

A recent evaluation of ADB's research investments through the international centers reported favorably on the achievements of the rice and maize biotechnology networks, and recommended further support for this type of research cooperation. The networks provide a platform by which countries and the IARCs can collaborate in the use of the new tools of functional genomics to identify genes that control important traits such as drought and salinity tolerance. This will speed up the breeding of crop varieties with these characteristics that have been difficult to address through conventional breeding. There will be increasing opportunities for cooperation between the networks, as more knowledge is gained on the commonality of genes between species and their control.

Australian Government

Over the last 15 years, ACIAR has entered into more than 100 biotechnology R&D partnerships with at least 10 Asian countries in support of more than 600 active or completed projects. The emphasis of this work has been on developing diagnostics and vaccines for a large suite of diseases of tropical livestock, with some recent work on fish and shrimp being undertaken. Most of these projects have been implemented through government programmes, but NGOs are becoming increasingly involved. Molecular marker methods for identifying disease-resistant genes and prolificacy in livestock have also been developed. Several projects on biotechnology for rumen manipulation have been carried out.

Crops and forestry work has focused on development of diagnostics for diseases (viral, fungal, mycoplasma, and bacterial) and contaminants in tropical crops, and the application of biofertilizers, bioremediation technology, and biofumigants. Molecular markers have been developed for the improvement of cereals and tree species. In cooperation with IRRI, attempts are being made to develop apomixis systems for rice. Tissue culture methods for micropropagation and conservation of several species, including sweetpotato, taro, tropical fruits, coconut, green tea, and tree species such as mangrove are being developed.

Eight of ACIAR's current or completed projects have included aspects of plant genetic engineering, with target crops being cereals and pulses, groundnut, and several tropical fruits. Target characters include virus resistance and quality defects related to ripening processes. These collaborative projects were developed at the request of the Asian countries, which fully approved regulatory procedures. ACIAR also provides core funding to many CGIAR and other IARCs, a proportion of which is applied to biotechnology R&D.

Australian support has been provided through AusAID for an Association of Southeast Asian Nations (ASEAN)/Australia regional biotechnology network, mainly concerned with food, microbial, and industrial biotechnology. AusAID also supports the characterization and conservation of genetic resources of taro and forest genetic resources in the South Pacific. AusAID has supported several biotechnology seminars. A major seminar in 1989 at the Australian Academy of Sciences on the potential of agricultural biotechnology in international development reported on the outcomes of a joint study cosponsored by AusAID, ACIAR, ISNAR, and the World Bank.

In the area of human resources, the Crawford Fund has sponsored several master classes in biotechnology for senior policymakers over the past decade.

Agency for International Development

The USAID is supporting applications of biotechnology through bilateral activities in several countries, including India, Indonesia, and Pakistan. USAID is also providing specific support for biotechnology applications within the programmes of IARCs. This includes support for research by IRRI on improving the nutritional quality of rice by increasing its vitamin A and iron content, and support for research by the Tata Energy Research Institute on the development of *golden* mustard.

World Bank

The World Bank has supported the development of infrastructure and human resource development for biotechnology in several Asian countries over the past 15 years. This support has come through loans in the agricultural sector, science and technology, and education. There are currently substantial components for biotechnology within agricultural technology projects in India, Indonesia, and Pakistan. The Inter University Centers for Biotechnology in Indonesia were built with a $150 million loan in the 1980s. The Republic of Korea also used World Bank loans to develop its infrastructure in biotechnology. The World Bank is also one of the main financial supporters of CGIAR centers. The centers invest about 10 percent of their total annual budget of $340 million in the applications of biotechnology.

The Rockefeller Foundation

Operational from 1984-1999, the Rice Biotechnology Network sponsored by the Rockefeller Foundation was very active and successful in the region. The program's two objectives were to (i) to create biotechnology applications to produce improved rice varieties suited to developing country needs, and (ii) to train scientists in developing countries to use the techniques and adopt them to their own objectives. A network of about 200 senior scientists and 300 trainee scientists from all the major rice producing countries of Asia and a number of industrialized countries participated. The programme was funded at approximately $5 million per year for 15 years.

Another outcome of the programme was the development of promising new technologies for the control of rice pests and diseases and improving the nutritional quality of rice. The work is being continued by national governments, ARBN, and IRRI to do the field evaluation and distribute new rice varieties in Asia. The Rockefeller Foundation presently concentrates its biotechnology programmes in Asia on drought tolerance in rice and maize. It is also examining innovative means of dealing with access to technologies and IP issues.

United Nations Agencies

The Food and Agriculture Organization (FAO) is giving high priority to biotechnology within its Asia/Pacific regional programmes. During 1989-1993, the United Nations Development Programme financed the establishment of biotechnology centers in eight countries (PRC, India, Indonesia, Republic of Korea, Malaysia, Pakistan, Philippines, and Thailand) to share rDNA techniques in animal improvement, embryo transfer, and disease control. In 1994 the United Nations Industrial Development Organization (UNIDO) established an International Centre for Genetic Engineering and Biotechnology in New Delhi, India, to assist Asian countries in the applications of biotechnology to important crops of the region.

RISK ASSESSMENT AND MANAGEMENT IN AGRICULTURE

Because of the inherent financial risks involved in agricultural production, farmers commonly insure their crops and livestock for potential physical risks such as hail, frost, drought, and disease. Insurance companies that underwrite these risks calculate the cost to purchase the insurance based on risk assessment tables that are developed by the insurer. Factors considered in determining the insurance rate include the physical risks, as well as the claim history on a specific parcel of land and the current management of the farm, to assess whether best practices are being used, thereby reducing risks.

PUBLIC PERCEPTION OF RISKS

As with any science and technology, biotechnology can bring with it benefits and risks. It is the risks of agricultural biotechnology that have received widespread publicity in the media even though biotechnology has also been applied to health and industrial sectors. Environmental NGOs have been particularly vocal in taking issue with the new technologies derived from or incorporating GMOs. As a consequence, in the public debate biotechnology has become synonymous with GMOs, although they are only one of the many products of biotechnology.

Curiously, biotechnology and GMOs in health-care products now in widespread use (insulin, hepatitis vaccine, medication for cardiovascular disease, etc.) or for industrial purposes such as bioremediation have elicited no such controversy. This can probably be attributed to the lack of benefits to consumers in the first generation of genetically modified (GM) crops. The main focus was on herbicide and insect resistance that benefited farmers, seed producers, and chemical companies. It is expected that the next generation of genetically modified foods will benefit consumers, nutritionally or from taste or storage benefits, and accordingly may be better accepted.

A number of food-related crises in recent years have made consumers particularly sensitive about food safety issues. Health and food safety concerns are again at the forefront in Europe following additional cases of mad cow disease *(bovine spongiform encephalopathy)* and the banning throughout the European Union of blood and bone meal in feed for all animals. These crises have not been caused by GMOs, but by the intensification of agriculture and food production, a fact that appears to have escaped public attention. In Europe in particular, demands have been made for informative food labelling so that consumers may, if they wish, avoid genetically modified foods.

The anti-GMO movement reveals profound mistrust of developments in science and technology and of the forces seen to be driving them. Genetically modified crops such as maize, sorghum, cotton, and canola have been widely grown for the last five years, yet no harmful effects on human health or the environment have been detected. That was one of the conclusions of the OECD-sponsored conference held in Edinburg, UK, in 2000. However, it is generally agreed that government and the private sector are responsible for monitoring the long-term effects of GMOs on human health and the environment.

POTENTIAL RISKS OF BIOTECHNOLOGY

The risks associated with modern biotechnology fall into four categories: food safety, environmental, socioeconomic, and ethical. Some of these concerns relate to potential risks inherent in modern biotechnology and can be described as technology-inherent. Others are related more to value systems or cultural practices and can be described as technology-transcending.

Food Safety Concerns

The potential risks of biotechnology on human health may include toxic reactions, increased cancer risks, food allergies, food contamination, and antibiotic resistance. There is also concern that GMOs in animal feed might present a health risk for consumers, or for the animal itself. Consumers are also concerned about the long-term health effects of genetically modified foods.

To address food safety concerns, the following safeguards have been adopted by some countries:

(i) Some countries have regulatory procedures, institutions, and infrastructure in place to ensure food safety. These regulations cover all aspects of the food chain, from farm inputs (including animal feed, feed additives, pesticides, fertilizers, veterinary drugs) through production and processing (including agricultural products, processed food, novel foods, food additives), to transportation, storage, and distribution.

(ii) Formal science-based procedures for risk analysis of food have been adopted by some countries. These continue to evolve with new scientific information about food safety, emerging pathogens, new technology, and consumer demands for a high level of public health protection. Generally, the procedures conform to international standards set by the Codex Alimentarius Commission of FAO.

(iii) Some countries have adopted a variety of approaches to regulate genetically modified foods, either by applying existing food safety measures or enacting new legislation.

(iv) In response to consumer demands, a growing number of countries are introducing labelling. Opinions differ on whether labelling should be mandatory or voluntary, as well as on acceptable tolerance levels and on the type of information to be used. Two labelling issues currently being addressed are segregation and traceability. Segregation relates to the ability to attest to the separation of genetically modified and non-genetically modified crops (as in the case of soybean in processed foods). Traceability means being able to attest to the origins of food products. Traceability is of particular importance in the current controversy related to mad cow disease. Low cost, quick tests are now available to determine the presence of genetically modified products in food.

Environmental Concerns

A number of biotechnology applications are not seen to present any new threats to the environment. That is the case with tissue culture, diagnostics, and market-selected plant breeding. On the other hand, there is fear of potential risks from the release of GMOs into the environment. The potential risks of GMOs on the environment may include increased pesticide residues, genetic pollution, damage to beneficial insects, creation of superweeds and superpests, creation of new viruses and bacteria, and genetic bioinvasion.

To address environmental concerns, some countries have adopted the following safeguards:

(i) Most current biotechnology applications in tissue culture and micropropagation, diagnostics and vaccines, and markerassisted plant breeding are subject to existing regulations. They include phytosanitary regulations and plant quarantine, varietal certification of seeds, and veterinary product regulations. These regulations usually conform to international standards, guidelines, or recommendations such as those set by the International Plant Protection Convention or the International Bureau of Epizootics for animal products.

(ii) A number of countries have introduced new requirements and procedures for the environmental release of GMOs. Procedures for hazard identification and risk assessment of GMOs are now well-established in most OECD countries, and in some Asian countries.

Socioeconomic Concerns

Modern biotechnology R&D has been conducted in an institutional and economic environment that differs significantly from the development of the earlier Green Revolution technologies. While the latter were essentially the prerogative of public research institutions and philanthropic foundations, developments in biotechnology have been driven essentially as a competitive, commercial endeavor in which powerful private sector actors compete.

The major socioeconomic risk of agricultural biotechnology stems from the fact that the research, development, commercialization, and distribution of new biotechnological products have been carried out mainly in developed countries by a few, large, multinational companies. These companies have focused on temperate crops for large farmers in developed countries. Undertaking R&D on Asia's basic food crops for small farmers in rainfed and marginal areas is of little interest because they see limited returns from such investments. If this trend continues, modern biotechnology will aggravate the income disparity between developed and developing countries, and between large and small farmers. Increased public investments in agricultural biotechnology are necessary to ensure that small, poor farmers have access to biotechnology. Governments must also address the potential gender disparity in the access to technology, and the negative impact specific to women, since no technology is gender neutral. Lessons from earlier agricultural technological changes should be used as a caveat.

Ethical Concerns

Ethical issues may stem from the uneven or inequitable impact of an expanding global economy; from the national social, economic, and politiRisks Assessment and Management 41 cal context; or from individual values. Indeed, it may be difficult to reconcile personal values with what a majority regard as the common good. One of the ethical concerns raised by biotechnology—and particularly GMOs—is that it is unnatural and an unwarranted tampering with nature. However, seen in historical perspective, most technology developments in agriculture over the centuries have involved, in one way or another, efforts to overcome the vagaries of nature.

RISK MANAGEMENT

To address the potential risks of biotechnology in agriculture, there is a need for Asian countries to establish effective biosafety procedures for the development, testing, and release of the technologies.

Emerging Regulatory Systems for Biotechnology

A growing number of international organizations have become involved in food and environmental safety. Concerted international efforts to develop agreed upon scientific approaches to biosafety date from 1975, when a group of scientists from the United States National Academy of Sciences expressed concern regarding the potential biological risks of rDNA molecules. Efforts to ensure the science-based, case-by-case risk identification, assessment, and management of GMOs have thus been pursued for some 25 years.

In the industrialized world, OECD has been instrumental in forging agreements on scientific principles regarding the safe applications of rDNA organisms in industry, agriculture, and the environment. Since the 1980s, OECD has established guidelines, first for (i) laboratory-based experimentation, then (ii) for small-and large-scale field trials of genetically modified plants and organisms, and finally (iii) for their commercialization and release into the environment or food chain.

During the 1990s, a growing number of international organizations including the United Nations Environmental Programme (UNEP), UNIDO, World Health Organization (WHO), and FAO have become involved in activities related to biosafety. Both UNIDO and UNEP have developed biosafety guidelines. FAO and the WHO are jointly responsible for the Codex Alimentarius Commission, which sets international standards for food safety. Codex has set up a new Intergovernmental Task Force on Foods Derived from Biotechnology to develop general principles for risk analysis and to provide guidance on risk assessment.

Other international organizations conducting work on the safe use of biotechnology include the International Centre for Genetic Engineering and Biotechnology, the United Nations Conference on Trade and Development, and, for many aspects of animal as well as human health, the International Bureau of Epizootics.

Cartagena Biosafety Protocol

Preoccupation with the protection of biological diversity, necessary for sustaining agriculture and food production and, indeed, life itself, became a topic of international debate during the 1980s. As a result, FAO established the International Undertaking on Plant Genetic Resources as a nonlegally-binding agreement for cooperation in the conservation of genetic material. The agreement was based on the universally accepted principle that plant genetic resources are a heritage of mankind and consequently should be available without restriction. The legally binding Convention on Biological Diversity, which came into force in 1993, encompasses not only plant genetic resources but all living organisms. In contrast to the FAO undertaking, the Convention affirms that States have sovereign rights over their own biological resources.

In January 2000, the legally binding Cartagena Protocol on Biosafety was adopted. This Protocol lays the foundation for a global system for assessing the impact of GMOs on biodiversity. It outlines the obligations of all countries that are signatories to protect biological diversity from the potential risks imposed by living modified organisms resulting from modern biotechnology. At the heart of the Protocol is the concept of advanced informed agreement to the import of GMOs. This means that an exporting country must inform an importing country of its intention to export. It must also provide information on the GMOs, including an appropriate risk analysis. For its part, an importing country is required to give its informed agreement to the import. This implies that, although signatory states have the right to prohibit the import of GMOs if advanced informed agreement has not been given, they must also be capable of determining the validity of a safety assessment.

In addition, the Protocol advocates the "precautionary approach," described in the 1992 Rio Declaration on Environment and Development as a principle of international environmental law. The Protocol also proposes to establish a Biosafety Clearing-House for collecting, sharing, and disseminating information on risk assessment and management, which is particularly important for developing countries. Thus far, 81 countries have signed the Protocol, including Bangladesh, PRC, Indonesia, Republic of Korea, Malaysia, Philippines, and Sri Lanka.

Approaches in Selected OECD Countries

A sample of industrialized countries shows a diversity of approaches to food and environmental safety and in the agencies or ministries that play key roles in the national biosafety system. For example, the US has chosen to develop biosafety capacity within existing institutions (notably the Environmental Protection Agency and the Department of Agriculture). Denmark has designated the National Forest and Nature Agency as the lead agency for biosafety. In Norway, in accordance with the Gene Technology Act, the application of GMOs is subject to ethical and social conditions. A GMO can be approved for use only after the assessment of its benefit to the community and its contribution to sustainable development. In Australia, apart from the different ministries involved, an Advisory Committee on Genetic Engineering was created some years ago.

That will soon be replaced by an office with an enforcement rather than an advisory role. A public hearing must be held before authorization is given to release a GMO. Australia has recently created the Office of the Gene Technology Regulator to operate in the Commonwealth Department of Health and Aged Care. Japan has based its risk assessment system on the principles and concepts developed by OECD. Different Japanese ministries and agencies have developed guidelines in accordance with their responsibilities on a product-sector basis. A growing number of countries are introducing guidelines or mandatory measures for labelling genetically modified foods. In the US, formerly strongly opposed to labelling, the Food and Drug Administration has recently been charged with developing guidelines for voluntary labelling of food products to indicate whether or not they contain GMOs. In the European Union, food products containing more than 1 percent GMO must be labelled, stating clearly: "this product contains GMOs." The Australia New Zealand Food Standards Council is drawing up a new Food Standards Code. Labelling requirements will be drawn up in accordance with this joint Code. Japan is planning the introduction of mandatory food labelling in April 2001.The OECD has also played an active role in promoting the harmonization of biosafety procedures among its member countries so that information and data gathered in the course of risk assessment may be mutually acceptable.

Emerging Regulatory Systems in Asia

A growing number have set up national biosafety systems; others are formulating national biosafety guidelines. Most countries have not yet completed guidelines for commercialization. The PRC is the only country in Asia where transgenic crops have been approved and released for sale. A number of countries are also in the process of introducing food labelling. Indonesia is planning to introduce mandatory labelling in April 2001.

ASEAN has undertaken a number of initiatives aimed at the harmonization of biosafety procedures among its member countries. In 2000 it also conducted a joint workshop with the Ministry of Science and Technology of the PRC to discuss issues related to transgenic plants. The Asia-Pacific Economic Cooperation has a number of capacitybuilding and technical cooperation activities in support of risk assessment, risk management, risk communication, and public acceptance. Responsibility rests with its Agricultural Technical Cooperation Experts Group and its Sub-Group on Research, Development and Extension of Agricultural Biotechnology.

Although some Asian countries have regulations in place, weak law enforcement sometimes results in the release of GMOs without proper procedures having been followed. For example, P.T. Monagro Kimia (a subsidiary of Monsanto) released Bt cotton in Indonesia without complying with Indonesian environmental law on safety procedures, environmental assessment, and information disclosure. Law enforcement in Asia needs strengthening to avoid risk to human and environmental health.

BIOSAFETY CAPACITY BUILDING IN ASIA

A growing number of initiatives are being taken to provide support for Asian countries in developing their capacities to safely manage biotechnology. A representative sample follows.

Following the signing of the Biosafety Protocol, UNEP, through its Global Environment Fund, is preparing a major capacity-building programme for developing countries. It will also be responsible for the establishment of the Biosafety Clearing-House. FAO serves as the secretariat for the International Plant Protection Convention, which has recently established a working group on the Phytosanitary Aspects of GMOs, Biosafety and Invasive Species. UNIDO has conducted training courses in biosafety in a number of Asian countries. In collaboration with OECD, it has compiled an information source and database (BIOBIN), which provides information on biosafety methods and procedures, and on the status of the regulatory frameworks in individual countries. OECD has established a network of international organizations with activities related to biosafety to enhance the exchange of information and to facilitate cooperation.

IBS provides a number of services to enhance developing country capacities to manage agricultural biotechnology. It gives short training courses annually for managers of agricultural biotechnology research programmes in Asia. Biosafety is a key element in these courses. IBS also provides advice to individual countries.

USAID is involved in a number of biosafety training activities in Asian countries, some of which have been provided by its Agricultural Biotechnology Services Program. ISAAA is currently setting up a Global Knowledge Centre on Crop Biotechnology based at ISAAA's SEAsia Centre at Los Baños in the Philippines. ISAAA also organizes capacity-building workshops and training for regulatory officials and scientists from the Asian region.

FUTURE CHALLENGES

The population of Asia is projected to increase from 3.0 billion to 4.5 billion in the next 25 years. During the same period, the urban population will nearly double from 1.2 billion to 2.0 billion, as rural people move to the cities in search of employment. These increases will place massive pressure on developing member countries (DMCs) of ADB to increase food production. Food demand is influenced by population growth, urbanization, income, and associated changes in dietary preferences. Urbanization and income growth frequently lead to shifts from a diet based on root crops (cassava, yam, and sweetpotato), sorghum, millets, and maize to rice and wheat, which require less preparation time, and to more meat, milk, fruits, vegetables, and processed foods. This dietary transition has already happened in much of the region. Meeting the food needs of Asia's growing and increasingly urbanized population requires increases in agricultural productivity and matching these increases to dietary changes and rising incomes.

To meet this demand, cereal production will need to be increased by at least 40 percent from the present level of about 650 million tons annually, most of which will have to come from yield increases. In addition, meat demand will double during the period. Production increases will have to be achieved by increasing yields in a sustainable way to conserve diminishing and degraded natural resources. Nearly all of these production increases will need to take place in DMCs themselves because on average 90 percent of the world's food is consumed in the country where it is produced. Food imports are not only expensive but discourage the creation of employment, which is badly needed in the rural areas.

In this millennium, we face a food, feed, and fiber production challenge in highly complex farming systems for several reasons:

(i) Water will become the most important limiting factor in agricultural production because the quality and quantity of water will decline as a result of pollution, forest degradation, and increased agricultural, domestic, and industrial use.

(ii) Urbanization will mean the loss of agricultural land to residential and industrial development, and a decline in the number of farm workers.

(iii) Most farmers are poor with small landholdings.

(iv) Farming systems are commonly heterogeneous with mixes of food crops, livestock, and trees.

(v) About 70 percent of the cultivated land is rainfed with unreliable distribution and intensity of rainfall.

Thus, the increase in food production during the next 25 years will have to be achieved using less labour, water, and cultivated land. This can be done only if scientists can develop new crop varieties with high yield potential and high water use efficiency. New understanding of plant and animal genes may offer ways to increase crop yields to the levels required to adequately and sustainably feed the growing population in Asia. Thus, developments in modern biotechnology could make extremely important contributions to future agricultural growth, food security, and poverty reduction. Increasing smallholder agriculture productivity will not only increase food supplies, but will reduce poverty and malnutrition, increase food access, and, improve living standards of the poor.

2

Agriculture and Poverty

INTRODUCTION

Agriculture can drive growth and development. The way it contributes – or not – to poverty reduction depends on many factors. Investment in agriculture can help promote food security, bring more work to the area, and reduce the proportion of their budget families spend on food. Governments can help by directing investments in ways that prioritize those benefits. With new threats to agriculture emerging as a result of climate change, extra investment is needed to establish more sustainable water management practices, to promote crop diversification and to encourage farmers to conduct their agricultural activities in other locations.

At the dawn of the twenty-first century, about 900 million or 68 percent of the world's poor people live in Asia; about 500 million in South Asia, 300 million in East Asia, and 100 million in Southeast Asia and the Pacific. In addition, about 526 million people, including 160 million children, are undernourished. Not only do they lack access to sufficient money to buy food and other essentials, but neither do they have access to sufficient schooling, adequate housing, nor medical care. Those in rural areas are often short of water and fuel. Fertile land and water for farming are increasingly scarce. Urban poor lack money to buy enough food. That which they can afford may be deficient in protein and essential vitamins and minerals.

Although the absolute numbers of people living in poverty in Asia today are unacceptable, the situation could be much worse. In 1970, 60 percent of all Asians lived in poverty. That figure has been cut by almost half, with about one third of all Asians living in poverty in 2000. Also, countries such as Bangladesh, the People's Republic of China (PRC), and India have moved from periodic famines to almost self-sufficiency in food production. However, further efforts are needed to reduce poverty by another 50 percent by 2015, as targeted by world leaders during the World Food Summit in 1997. The latter half of the twentieth century saw impressive advances in science and technology. We now have the capacity to apply this knowledge to reduce poverty and improve food security. This Working Paper discusses how biotechnology can be used to safely and effectively reduce poverty and improve food security in Asia.

AGRICULTURAL PRODUCTIVITY AND TARGETED ANTI-POVERTY PROGRAMS

While efforts to increase agricultural productivity and thereby increase farm incomes and employment are a major instrument for poverty alleviation, they will need to be supplemented by special targeted programme aimed at improving the welfare of vulnerable groups in rural areas.

Employment programmes in rural areas have been the most important of these anti poverty programmes and India has a long history of such programmes. Building on this tradition, a Rural Employment Guarantee Act has been enacted which provides assurance of up to 100 days of employment at the minimum wage to each household in rural areas wishing to make use of it.

The employment would be provided on projects chosen by the elected village councils and the guidelines specify that top priority should be given to irrigation and water management schemes. Unlike earlier employment programmes, this scheme includes a guarantee in the sense that if employment cannot be provided, unemployment compensation will be provided at least 25 percent of the wage. Although the programme open to each household, actual demand for employment is expected to be limited to households below the poverty line. The coverage of the Act will initially be implemented to 200 of the most backward districts (about one-third of the total districts in the country). Together with other special programmes relating to provision of housing for the poor, old age insurance, and schemes for supporting self employment, this programme will provide an element of social security that should help to reduce poverty.

REDUCING POVERTY THROUGH AGRICULTURAL SCIENCE

Science and technology underpinned the economic and social gains in Asia over the past 30 years. In agriculture, these gains came to be known as the Green Revolution. Between 1970 and 1995, cereal produc-tion in Asia doubled, calorie availability per person increased by 24 percent, and real food prices halved. Although the region's population grew by 1 billion people, overall food production more than kept pace with population growth. These food production increases were achieved largely by the cultivation of high-yielding varieties (HYVs) of rice and wheat, accompanied by expansion of irrigated areas, increases in fertilizer and pesticide use, and greater availability of credit.

The scientific basis for the Green Revolution stemmed from national and international research programmes that led to the development and distribution of new HYVs, particularly of rice and wheat. The first generation of these new varieties were based on the introduction of new genes for dwarfing that made the HYVs shorter, more responsive to fertilizers, and less prone to falling over or lodging when fertilized and irrigated. Subsequent varieties also carried genes that gave increased pest and disease resistance and improved taste and grain quality.

The key elements in improving food security in Asia from 1970-95 were government policies reflecting a belief that investments in increasing agricultural productivity were a prerequisite to economic development. These national policies were supported by political leaders in Asia and by both the public and private sectors of the international community. This mix of supportive public policies, scientific discoveries, and public and private investments in rural Asia, particularly in irrigation, credit, and inputs, led to substantial reductions in poverty and improved food security throughout Asia over the past 30 years. Increased agricultural productivity, rapid industrial growth, and expansion of the nonfarm rural economy have all contributed to almost a tripling of per capita gross domestic product across Asia since 1970.

PRESENT PROBLEMS

Despite these successes, problems remain. The intensification of agriculture and the reliance on irrigation and chemical inputs has led to environmental degradation, increased salinity, and pesticide misuse. Deforestation, overgrazing, and overfishing also threaten the sustainable use of natural resources.

Green Revolution technologies had little impact on the millions of smallholders living in rainfed and marginal areas, where poverty is concentrated. Furthermore, the Green Revolution has already run its course in much of Asia. Wheat and rice yields in the major growing areas of Asia have been stagnant or declining for the past decade, while population continues to increase. The key lessons learned from the Green Revolution are: (i) it has benefited farmers in irrigated areas much more than farmers in rainfed areas thus worsening the income disparity between the two groups, (ii) it overlooked the rights of women to also benefit from the technological advances, and (iii) it promoted an excessive of use of pesticides that are harmful to the environment. As countries became self sufficient in food, government investments declined in the agricultural sector and in science and technology across the region. This reflects a worldwide trend towards declining public investments in the rural sector and in agricultural research and development (R&D), nationally and internationally.

In Asia, private sector investments in the rural sector and related R&D have concentrated on export commodities. The downward trends in public investments by governments and development agencies in smallholder agriculture over the past decade have not been matched by a concomitant rise in private investments. Similarly, there is little (and few incentives for) private R&D on the food crops, livestock, fisheries, and aquaculture systems important for food security and poverty reduction in rural Asia.

RATIONALE FOR ADB'S INVOLVEMENT IN BIOTECHNOLOGY

Biotechnology has the potential to (i) increase crop and animal productivity; (ii) improve nutritional quality; (iii) broaden tolerance of crops for drought, salinity, and other abiotic stresses; and (iv) increase resistance of crops to pests and diseases. These potential benefits will have significant impact in increasing food production and reducing poverty in DMCs if they can be applied to problems of the poor farmers in the tropics. As with any technology, biotechnology brings with it potential risks. To maximize the benefits and minimize the risks, the introduction and use of biotechnology in DMCs or elsewhere must be thoughtfully managed by the public and private sectors. The key risks that relate to the application of new developments in biotechnology for the public good are food and environmental safety, economic concentration, and intellectual property (IP) management.

To explore the opportunities and risks of biotechnology, ADB has decided to undertake a comprehensive study on the use of biotechnology to reduce poverty and achieve food security in cooperation with the Australian Centre for International Agricultural Research (ACIAR)/ Australian Agency for Agricultural Development (AusAID). There are many reasons to support this initiative:

(i) As a premiere development institution, ADB should be well informed on the development of biotechnology and the issues surrounding modern biotechnology.

(ii) Biotechnology has been identified by the Rural Asia Study as one of the emerging challenges in Asia.

(iii) Biotechnology has been identified by the Consultative Group on International Agricultural Research (CGIAR) as a powerful tool to fight poverty when used in conjunction with other agricultural research.

(iv) ADB is responsible for assisting the DMCs in dealing with potential risks of biotechnology and providing information on various biotechnology issues.

(v) ADB's Policy on Agriculture and Natural Resources Research recommends that ADB support research to increase farm productivity through modern biotechnology approaches, while giving appropriate consideration to indigenous and traditional knowledge.

ADB'S POVERTY REDUCTION STRATEGY

While the need for further intensification of agricultural production in Asia is clear, intensification strategies must change to avoid adverse environmental impact and to reverse the effects of earlier practices. Enhanced, but inefficient use of irrigation and mineral fertilizers over the past three decades has had negative side effects such as soil salinity and nutrient leaching. With crop intensification, incidences of pests and diseases also increased. The following strategies are needed to meet the food demand in Asia over the next 25 years.

- Sustainable productivity increases in food, feed, and fiber crops.
- Reducing chemical inputs of fertilizers and pesticides and replacing these with biologically based products.
- Integrating soil, water, and nutrient management.
- Improving the nutrition and productivity of livestock and controlling livestock diseases.
- Sustainable increases in livestock, fisheries, and aquaculture production.
- Increasing trade and competitiveness in global markets.

As rural poverty persists in Asia, agriculture will play a prominent role in achieving equitable and sustainable rural growth in the twenty-first century. Even when rural people do not work directly in agriculture, they rely on nonfarm employment and income closely related to agriculture.

Where there are large numbers of rural poor, agricultural growth is a catalyst for broad-based economic growth and development. Agriculture's linkages to the nonfarm economy generate employment, income, and growth in the rest of the economy.

A healthy agricultural economy also offers incentives for natural resource conservation.

The ADB Poverty Reduction Strategy sees three factors, pro-poor, sustainable economic growth; good governance; and social development as key elements for reducing poverty. Biotechnology may contribute towards achieving the poverty reduction goals in several components of the strategy:

- Increasing priority for agriculture and rural development.
- Addressing environmental considerations.
- Increasing the public and private sector roles in poverty reduction.
- Encouraging regional and subregional cooperation.
- Coordinating efforts of funding agencies.

The challenge is how to use new scientific developments such as biotechnology, together with information and communications technology, to make the complex agricultural systems of Asia more productive and sustainable. Good governance is also crucial to ensure that new agricultural biotechnology reaches the poor.

SCIENTIFIC CONTEXT

PROMETHEAN SCIENCE

The pace of change in modern science has led some to term it Promethean science, acknowledging both its risks and benefits. Modern science encompasses new developments in the biological, physical, and social sciences. In biology, discoveries over the past 20 years allow the better understanding of the structure and function of human, animal, and plant genes. At the same time, new discoveries in the physical sciences underpin the revolution in information and communications technologies. Geographic information systems enable characterization of agro-ecosystems and offer means by which new technologies can be customized to the needs of particular agro-ecosystems. The biological and physical sciences also interact in new ways. For example, the ability to analyze large volumes of data is a critical component of various genome projects that are mapping all the genes in an organism, as in the Human Genome Project.

New developments in the social sciences underpin community participation in technology development and evaluation (sometimes termed agro-ecological methods). Participatory methods developed in the social sciences can help in understanding problems and the researchable issues, particularly those of small farmers operating in marginal environments. They may also be used to clarify the concerns of rural and urban dwellers in regard to the deployment of new technologies, including the products of biotechnology. Integration of all branches of modern science and traditional knowledge is required to develop knowledge-intensive solutions to the problems of rural Asia. These solutions need not be only technically feasible but also socially acceptable. Indeed, the potential value of modern science to agriculture and the environment in Asia will require the efforts of all stakeholders, including civil society, farmer cooperatives, producers, consumers, governments, and development agencies.

DEFINITION, HISTORY, AND SCOPE OF BIOTECHNOLOGY

Biotechnology, broadly defined, includes any technique that uses living organisms, or parts of such organisms, to make or modify products, to improve plants or animals, or to develop microorganisms for specific use. It ranges from traditional biotechnology to the most advanced modern biotechnology. Biotechnology is not a separate science but rather a mix of disciplines (genetics, molecular biology, biochemistry, embryology, and cell biology) converted into productive processes by linking them with such practical disciplines as chemical engineering, information technology, and robotics. Modern biotechnology should be seen as an integration of new techniques with the well-established approaches of traditional biotechnology such as plant and animal breeding, food production, fermentation products and processes, and production of pharmaceuticals and fertilizers.

The key components of modern biotechnology are listed below.

- Genomics: The molecular characterization of all genes in a species.
- Bioinformatics: The assembly of data from genomic analysis into accessible forms, involving the application of information technology to analyze and manage large data sets resulting from gene sequencing or related techniques.
- Transformation: The introduction of one or more genes conferring potentially useful traits into plants, livestock, fish and tree species.
- Genetically improved organism.
- Genetically modified organism (GMO).
- Living modified organism (LMO).
- Molecular breeding: Identification and evaluation of useful traits in breeding programmes by the use of marker-assisted selection (MAS);
- Diagnostics: The use of molecular characterization to provide more accurate and quicker identification of pathogens; and
- Vaccine technology: The use of modern immunology to develop recombinant deoxyribonucleic acid (rDNA) vaccines for improved control of livestock and fish diseases.

Biotechnology consists of a gradient of technologies, ranging from the long-established and widely used techniques of traditional biotechnology to novel and continuously evolving modem biotechnology techniques. During the 1970s scientists developed new methods for precise recombination of portions of deoxyribonucleic acid (DNA), the biochemical material in all living cells that governs inherited characteristics, and for transferring portions of DNA from one organism to another. This set of enabling techniques is referred to as rDNA technology or genetic engineering.

Modern biotechnology presently includes the various uses of new techniques of rDNA technology, monoclonal and polyclonal antibodies, and new cell and tissue culture methods. Over the past two decades the number of significant advances in modern biotechnology for understanding and modifying the genetics of living organisms has increased dramatically. That has led to greatly increased interest and investment in biotechnology, and increasing concerns as to the power of the new technologies and their safety.

ECONOMIC CONCENTRATION IN AGRICULTURAL BIOTECHNOLOGY

Modern biotechnology R&D has been conducted in an institutional and economic environment that differs significantly from the development of the earlier Green Revolution technologies. While the latter were essentially the prerogative of public research institutions and philanthropic foundations, the application of modern biotechnology to agriculture is essentially a competitive, commercial endeavor in which powerful private sector interests compete. Similarly, while the Green Revolution technologies were essentially dedicated to the public, the strengthening and extension of IP protection, particularly since the conclusion of the Uruguay Round of trade negotiations has increased the private character of biotechnologies. Multinational companies in the seed, agricultural chemical, pharmaceutical, and food-processing industries play a major role in biotechnology research. They have invested heavily in in-house research facilities, commissioned research, taken equity positions in new biotechnology firms, and entered into contractual arrangements with public research institutions or universities. As a result of mergers and acquisitions in the past few years, the development of new biotechnology applications in agriculture has become increasingly concentrated in the hands of a decreasing number of companies. The dominant companies that operate within global markets are Aventis, AgrEvo, Dow, DuPont, Monsanto, and Syngenta.

Biotechnology R&D has been concentrated in a limited number of industrialized countries, with the United States (US) in the lead in financial and human resources. A growing number of developing countries have invested in biotechnology R&D, but the amounts are small compared to the sums invested by private companies in the industrial world. While private sector investment in agricultural research in general is increasing in developed countries, there is still little private sector biotechnology research effort in developing countries, particularly in Asia.

The commercialization and distribution of new agricultural biotechnology products, particularly transgenic crops, is also concentrated in Organisation for Economic Co-operation and Development (OECD) member countries, with a few exceptions. These products are for the most part crops of economic importance in industrial country agricultural production and in world trade in agricultural commodities, mainly soybean, maize, cotton, and canola. During the past five years, the area under GMOs3 has increased rapidly from 1.7 million ha to 44.4 million ha, 75 percent of which are in the US. The remaining 25 percent are distributed in both developed and developing countries, including Argentina, Australia, Bulgaria, Canada, PRC, France, Germany, Mexico, Portugal, Romania, South Africa, Spain, Ukraine, and Uruguay.

Relatively little biotechnology research is being undertaken on the problems of small farmers in rainfed and marginal lands. Neither is there much interest in Asia's basic food crops: rice, tropical maize, wheat, sorghum, millet, banana, cassava, groundnut, oilseed, potato, sweetpotato, and soybean. These are considered *orphan crops* because of the private sector's reluctance to work on them. That focus is unlikely to change because of the perception that investments in such orphan crops and from working on problems of small farmers yield limited returns. To participate more fully in the biotechnology revolution, Asian governments will need to expand their capacities to undertake biotechnology research linked to the problems of small farmers and orphan crops. In certain situations, however, there may be opportunities to purchase, license, or import technology applicable in Asia.

APPLICATIONS OF BIOTECHNOLOGY

The applications developed from the new methods in biotechnology place them within the continuum of techniques used throughout human history in industry, agriculture, and food processing. Thus, while modern biotechnology provides powerful new tools, they are used to generate products that fill similar roles to those produced with more traditional methods.

There is now increasing use of modern molecular genetics for genetic mapping and MAS as aids to more precise and rapid development of new strains of improved crops, livestock, fish, and trees. Other biotechnology applications such as tissue culture and micropropagation are being used for the rapid multiplication of horticultural crops and trees. New diagnostics and vaccines are being widely adopted for the diagnosis, prevention, and control of fish and livestock diseases.

The science of genomics (the molecular characterization of all the genes in a species) has dramatically increased knowledge of plant genes and their functions. The new technologies enable greatly increased efficiency of selection for useful genes, based on knowledge of the biology of the organism and the role of specific genes in regulating particular traits. This will enable more precise selection of improved strains. These techniques may be used for more efficient selection in conventional breeding programmes. They may also be used for the identification of genes suitable for use in the development of transgenic crops. Thus far, scientists have completed genomic study on rice through the cooperative efforts of several international and private sector institutions led by Japan.

Modem biotechnology permits increased precision in the use of new techniques and a shorter time to produce results. For example, plant breeders and molecular biologists can collaborate to transfer to a highly developed crop variety one or two specific genes to impart a new character such as a specific kind of pest resistance.

New techniques of modern biotechnology accelerate plant and animal breeding. They offer possible solutions to previously intractable problems and difficult targets such as drought tolerance, and enable the development of new products. These products may include more nutritious food, crop varieties with improved tolerance for pests and diseases, and animal vaccines.

It is important to provide appropriate regulatory mechanisms to ensure that products produced by modern biotechnology are as safe as the products of traditional biotechnology. That is especially so when the products are GMOs that might interact with the environment. Our knowledge of genes is not matched by our knowledge of the gene-environment interaction or potential impacts of biotechnology on the environment. However, many of the Green Revolution technologies were also introduced without such understanding. At present, there is widespread distrust of biotechnology and the public needs to be engaged in dialogue before the technology is disseminated widely.

GENETICALLY MODIFIED ORGANISMS: FOOD AND ENVIRONMENTAL ISSUES

In industrialized countries, considerable progress has been made in methods, approaches, and experience in the safe management of GMOs. Overall these methods have proven to be effective, particularly with respect to food safety. The development of a transgenic soybean containing a protein derived from Brazil nuts, potentially useful in animal feed, was abandoned when the safety assessment revealed that the protein was probably an allergen. In response to consumer concerns, antibiotic markers are now being phased out as alternatives are being developed. That despite consensus among national and international regulatory authorities and the scientific community that the markers pose no threat to human health.

Despite acknowledged progress, areas of scientific uncertainty or disagreement persist. New challenges for risk/safety assessment, management, and monitoring will arise as a second generation of food, agricultural, and public health products emerges during this new decade. Not only are the ranges of organisms and numbers of traits expected to expand, so are the numbers and diversity of geographic and ecological sites in which they are released. These new products will not lend themselves easily to current approaches and methods to risk and safety assessment and to management. This highlights the need for constant review and improvement of assessment principles and procedures, and the urgent need for collection and analysis of ecological data. Other areas where consensus has not been reached among scientists themselves, or among scientists and policymakers, include (i) the precautionary approach as a method of dealing with scientific uncertainty, and (ii) methods for traceability.

The potential for the development of insect resistance to Bt transgenic crops is another area where the scientific community, industry, and environmental NGOs have not reached consensus. Risk management approaches at the farm level have so far been applied mainly on large-scale, commercial farms where monoculture is practiced. The same approaches may not be effective and, indeed, may not be necessary in the mixed farming systems of Asia.

Efforts to address the issue of resistance to transgenic crops have recently been initiated in Asia. A first International Consultative Workshop on Effective and Sustainable Use of Agricultural Biotechnology in Integrated Pest

Management in Developing Countries was held in November 2000. Organized by IRRI, Zhejiang University, and the International Organization for Biological Control, the workshop tried to determine strategies for the safe deployment of crops such as cotton and rice with novel Bt genes for pest resistance.

Among OECD countries, the continued monitoring of GMOs after release into the natural environment is also an area where opinions differ. On the one hand, it is argued that monitoring should be continued so that unforeseen risk can be managed and ecological impacts can be assessed.

On the other hand, it can be argued that the high costs of monitoring are not justified when transgenics have already successfully passed the hurdles of risk identification, safety assessment, and risk management.

It is clear that the potential risks of GMOs may vary by site and over time. Continued research will be necessary to improve data collection and analysis and to modify or devise new risk management strategies as needed.

FUTURE PERSPECTIVE

To date, a number of Asian countries have made substantial investments in biotechnology R&D initiatives. Some public research institutions have reached the stage of field testing of new crop varieties, major research projects are coming to fruition in IARCs in the region, and private companies are also conducting research in the region, or have products waiting to be commercialized and exported to the region. It will be necessary to have biosafety procedures in place to ensure that the benefits of these technologies are realized.

Although many opportunities exist, from donor and other development agencies, in support of capacity building for the safe management of biotechnology, and in setting up national systems, it is the responsibility of national governments to ensure that national regulatory systems are applied, enforced, and monitored. Setting up an effective national regulatory system for the safe management of biotechnology requires substantial resources, both financial and human. It is also a political process and, as such, requires political commitment.

The Cartagena Biosafety Protocol confers both rights and obligations on countries that are signatories. A country is free not to allow the import of GMOs on its territory. At the same time, it has an obligation to establish a national biosafety system.

This means that even if a country decides to ban the import or use of GMOs on its national territory, it will still need to comply with the obligation to develop a biosafety system for the control of such organisms.

It is anticipated that UNEP, the World Bank, and United Nations Development Programme (UNDP), through the Global Environment Fund, will contribute funding and technical assistance to developing countries for capacity building in biosafety. Even so, compliance with both the provisions of the Convention on Biological Diversity and the Cartagena Biosafety Protocol will place a heavy burden on some countries.

Apart from obligations under the Biosafety Protocol, if countries wish to control the import or introduction of GMOs on national territory, it will be necessary to have the means of detection in place. Similarly, from the perspective of international trade, when countries export agricultural or food products to countries that either prohibit the import of GMOs or require mandatory labeling for food or food products containing GMOs, the requisite regulatory procedures need to be in place.

3

Agricultural Growth and Performance: An Economy-wide Analysis

INDIA'S ECONOMIC REFORMS IN AGRICULTURE

Although India's economic reforms were initiated in June 1991, the process of liberalization was implemented gradually and thus it is difficult to assess the full impact of the liberalization measures. Nevertheless, an attempt is made to discuss what is observable in terms of agricultural growth.

One observation is that the expected increase in exports due to liberalization simply did not occur. India's share in world exports was 0.6 per cent in 1997; India has to aim for at least 4 per cent by 2005 in order to meet the growing import demands for capital goods, raw materials and crude oil as well as to meet her external financial commitments. For the last decade or so, India's share in world exports of agriculture has been between 2 per cent and 3 per cent. India is not as competitive as the other countries and calculations show that India's crop yields have increased at a slower rate over the 1990s.

In addition, the agricultural sector's output growth decreased to 2.9 per cent during 1992-1993 to 1998-1999. Kalirajan and others (2001) explain that two important reasons for the slowdown are that there was no major breakthrough in developing new high-yielding varieties during the 1990s and there was a decline in the environmental quality of land which reduced the marginal productivity of the modern inputs. What could this mean in terms of the effectiveness of the policies of reduced protection to industry, a market determined exchange rate and the opening of the agricultural sector to foreign trade?

First, although the reduction to protection of industry is substantial, there is reason to believe that the reduction was not necessarily sufficient to benefit the agricultural sector whose tariffs were also drastically reduced. Hence, the expected shift in resources to agriculture did not occur. Second, is the apparent ineffectiveness of the market determined exchange rate in boosting exports. This is however not surprising as the exchange rate may not be a key factor determining agricultural export demand for India. In general, unlike manufacturing industries, agriculture did not benefit much from these two policies because the share of imported inputs in the value of agricultural production is small. It is likely that a change in the mindset and attitude of farmers has yet to take place and there are delays or hesitation in embracing India's openness.

Third, in opening up the agricultural sector to foreign trade, India has taken major steps towards trade liberalization since 1991, partly on its own initiative and partly from its commitments to WTO. Kalirajan and others (2001) provide a detailed review of these reform procedures. But why have the benefits from trade liberalization been slow to come?

One reason is that prospects for growth in agricultural exports depend partly on domestic policies and partly on the removal of protectionist policies pursued by developed countries such as Japan and members of the European Union (EU). An OECD report (1998) estimated that the producer equivalent subsidy in the OECD countries increased by US$ 9.3 billion from 1988 to 1993 and this subsidy as a percentage of the value of production in 1997 was 9 per cent in Australia, 20 per cent in Canada, 47 per cent in EU and 70 per cent in Japan. These protectionist practices do not seem likely to come to an early end. An UNCTAD report (1999) noted that 29 member countries of the OECD spent an average of US$ 350 billion a year in agricultural support between 1996-98. Schumacher (2000) further reports that the EU provides product-specific trade distorting domestic support to at least 50 different agricultural products. The implication of these reports is that food exports from India may not show a large increase given the international environment and the still-existing restrictions on exports in the major importing markets based on the self-sufficiency argument and food security. Other macroeconomic factors, such as the recession in developed countries in 1996-98 as well as the 1997 South-East Asian financial crisis, have clouded the possibilities of increasing Indian exports.

Another problem faced by Indian agricultural exporters is the protectionist measures in the form of non-trade barriers that developed countries use to restrict market access. This is by tightening requirements of quality, testing and labeling, and anti-dumping and countervailing measures. For example, in May 1997, the EU banned marine products from India citing unhygienic processing conditions. The extra costs of meeting the standards required in export markets as well as costs associated with changes in the production mix and transactions associated with exports may well be discouraging Indian exporters.

One existing problem of India's agricultural protection is the use of input subsidies. The general argument favouring this has been that it is necessary to encourage the use of particular inputs for production for various benefits. For India, Gulati and Sharma (1995) show that the input subsidy in per cent of GDP increased from 2.13 in the triennium ending 1982-1983 to 2.73 in the triennium ending 1992-1993.

But the benefits of these subsidies have accrued to only certain classes of farmers in some regions cultivating irrigated crops. Furthermore, highly subsidized prices of inputs such as irrigation water and electricity for pump sets have encouraged cultivation of water-intensive crops, over-use of water, ground water depletion/salinity and water logging in many areas. Subsidy for nitrogen fertilizer on the other hand has resulted in nitrogen phosphorous potassium imbalance and acted as a disincentive for use of the environmentally friendly organic manure. As a result, the linkage between food crops and non-food crops, which include fodder, has been reduced.

These adverse consequences are a drain on the fiscal burden of central and state Governments. Thus, if not properly monitored, input subsidies can be counterproductive and, in this context, protection to lower costs of production should be done selectively in the course of liberalization.

In fact, Agenda 21 of the United Nations Conference on Environment and Development in 1992 stressed that there is a need for integration of environmental considerations in the pricing of natural and other resources in such a way that prices reflect social costs.

Such a pricing policy will not only lead to a more efficient use of scarce resources but also result in subsidy reductions and improvements in environmental quality. The money saved from the reduction of subsidies can be spent in the development of rural infrastructures, agricultural research, farmers' education and other forms of support for agriculture.

AGRICULTURAL GROWTH AND PERFORMANCE

While the above analysis has provided a general view of the impact of economic reforms, this chaper examines agricultural growth and performance in the states of Bihar, Karnataka, Tamil Nadu and Punjab with their attendant policy implications.

The yields for various crops in these states differ greatly. While Tamil Nadu had the highest yield in rice, oil seeds and sugarcane, Punjab enjoyed the highest yields in wheat, coarse cereals, pulses and food grains. Karnataka on the other hand is seen to do well in cotton and Bihar performed quite well in pulses and coarse cereals. Further analysis and findings by Kalirajan and others (2001) show that Punjab had made remarkable achievements on the agricultural front while Bihar had remained stagnant in the last two decades, with Karnataka and Tamil Nadu showing moderate achievement. Clearly, differences in physical endowments, climatic conditions and institutional characteristics are some of the reasons for the varying productivity performance. Thus, having across the board economic reforms is likely to work less effectively than state-specific policy measures that enable each state's agricultural yields to reach their full potential. The comparative advantage of each state's agricultural production should be determined and with inter-state restrictions removed, total agricultural output would see a very significant increase.

For example, Karnataka with less favourable soil and water resources should be given incentives to concentrate on agro-processed products and corporate agriculture in horticulture, floriculture and animal husbandry, or to undertake watershed development to help with dry land agriculture. Many studies have indicated that with watershed areas, productivity growth has been mainly due to seed and fertilizer use. Thus, this state has to be given input subsidies for high yielding seed varieties but at the same time, the farmers need to be educated on the over use of chemical fertilizers.

With Bihar, agricultural performance is problematic on many fronts. First, although demographic pressure has increased and agricultural technology has improved, most of the uncultivated land is concentrated in southern Bihar, where irrigation facilities have not kept pace and the soil is of poor quality. Given the physiography of southern Bihar, wells are also unsuitable and thus the dominant mode of irrigation has been through tanks whose expansion and maintenance has been neglected. Second, the infrastructural facilities of Bihar have been lagging as seen by the infrastructure development. Due to infrastructural bottlenecks, availability of modern goods and services has not increased or their supply remains costly or unreliable.

Third, agriculture in Bihar is dominated by small and marginal farmers and the prevalence of mass poverty is largely related to the backwardness of agriculture. Fourth and importantly, the state agricultural policies in Bihar are in dire need of review. The semi-feudal production condition still exists in rural areas and the ineffective protection of tenancy rights has hindered agricultural growth. The slow pace of land consolidation reflects inadequate financial outlays and a shortage of manpower. Kalirajan and others (2001) note that marketing and extension services in Bihar are also rather weak compared to the other states.

Punjab on the other hand, was one of the few states which enjoyed the success of land reforms and the high priority of investment in rural infrastructure. Also, the irrigation base of the small and medium sized farms was comparable to that of large farms. In addition, the Punjab Agricultural University at Ludhiana contributed to the development of new seed varieties.

However, there are clear signs of a decline in crop yields since the 1990s and this has been associated with the increasing use of fertilizers and excessive water use which have increased the unit cost of production as a result of declining soil quality. Hence, care is needed when providing further input subsidies in fertilizer and water use. Another related fact is the steep increase in wages in Punjab and in the absence of productivity increases, the cost increase has affected the profitability of farmers. With Tamil Nadu, the main crop has been rice as this state is blessed with two monsoons. But from 1992-1997, there has been a steady decline in the areas irrigated by canals and an increase in well-irrigated areas while the use of tanks remains an unreliable source of irrigation. However, major improvements in about 10 rice varieties released in the early 1990s can be expected to improve productivity growth in rice production although pests and diseases as well as imbalance in the use of fertilizers are major constraints. Thus Tamil Nadu could do with subsidies of pesticides and farmers should be educated on the more effective use of fertilizers to obtain high yields. Interestingly, the cropping pattern of late has shown increasing substitution of food crops by commercial crops but there is concern that the benefits will reach farmers only with the development of adequate infrastructure such as roads and markets. However, that Tamil Nadu has a higher index than the all India average of infrastructure.

Challenges of Globalization

It is important to realize that globalization poses many challenges to a developing country like India, which had relied on a state directed and regulated policy regime for more than four decades. In moving to a more open, market-based economy there are many transitional problems that the country has to manage. The Government must play a pro-active role in facilitating the globalization process so that the opportunity sets for the economic agents are widened and the adverse effects of globalization are minimized. The Indian Government must also prepare the necessary information base and develop its capacity to articulate India's concerns and policy trade-offs in the international forums for multilateral trade and environmental negotiations.

In addition, the Government should embark on an extensive programme to educate farmers on the need to meet the standards required in the export markets. In fact, India needs to seek technical assistance in creating the capacity for meeting such standards and to consider watershed developments for environmental considerations. Equally important is the need to disseminate information about possible export markets to farmers, so that market access is achieved at minimum cost. Given the requisite information about markets and profitability, the likelihood of farmers investing in post-harvest and processing technologies and storage and efficient transportation arrangements as well as developing supporting infrastructure is very high.

Although the brave and bold move by India to reduce the tariff rate for agricultural products from 113 per cent in 1990-1991 to 26 per cent in 1997-1998 deserves to be applauded, the question of whether India is ready to compete in world markets remains to be seen. The infant industry argument may still hold for India to shield itself from external competition but one can easily question the length of time that is required to that end. Also, a delay in opening up to foreign trade has the danger that local producers may become too complacent and never be ready for competition.

As India opens up externally, it is also expected to face vulnerability in the wider international price fluctuations and thus Acharya (1998) claims that a minimum price support scheme is important. These prices can also act as a signal to adopt modern inputs and invest in yield-raising infrastructure for increasing production. For instance, keeping basic staple food grains at reasonable prices would induce farmers to switch over to high value crops. However, during the 1990s, Kalirajan and others (2001) shows that procurement prices especially for rice and wheat have been increasing faster than the general price level. Such high prices along with guaranteed purchases by the Food Corporation of India have pushed up market prices. These higher prices are partly responsible for the large buffer stocks with the Food Corporation. If this trend continues, India's comparative advantage will be eroded.

With openness and high price instability, unstable export revenue can also be expected. One way of reducing such risk is for India to diversify her agricultural exports. For example, since 1990, even in commodities such as tea, coffee, cocoa and spices, where India is supposed to have a comparative advantage international prices have been unstable. Besides increasing the type of exports to obtain more export revenue, India should also seriously consider exporting more value added agricultural products through agro-processing such as processed vegetables, fruits, fish and meat products given that export or even local demand for basic agricultural products would decline as incomes rise. The move to higher value added activities within the agricultural sector also spells greater opportunities for industrialization and vice versa as borne by Kalirajan and Shand's (1997) findings of a bi-directional relationship between agriculture and industry for most Indian states. On the other hand, Sivakumar and others (1999) establish empirical evidence of high forward linkages of agriculture due to the presence of agro-industries while Satyasai and Viswanathan (1999) show the significance of the spillover effects to the industrial sector via the intensive use of purchased inputs in the agricultural sector.

The lack or slow pace of internal or domestic liberalization is also seen to hinder the possible gains from external or trade liberalization. For example, although central zoning restrictions have been abolished, state government restrictions on inter-state and even inter-district restrictions on marketing and movement of goods still exist in many cases. This interferes with the benefits from crop specialization and economies of scale arising from comparative advantage. The land market is another example of distortion whereby land ceilings exist preventing the operation of large-sized farms. This has led to the emergence of a large number of small economically unviable land holdings. The easy leasing of land should be permitted with assurance of resumption.

Yet another problem lies with the insufficiency of credit to agriculture. From 1995-1996, the Rural Infrastructure Development Fund was set up to allocate funds for the completion of projects and the government has committed itself to strengthening the cooperative credit structure through substantial refinancing and restructuring of the Regional Rural Banks. However, as mentioned earlier, due to varying institutional factors in the Indian states, these domestic reforms can be expected to yield quite different results.

Although India missed the opportunity to open up two decades ago, its attempts to do so now must be regarded as better late than never. Others such as Desai (1999) observe that, "the logic of the global economy as well as India's interests dictate that India become proactive in its liberalization policies. India must liberalize not because it has no choice but because it is the best choice".

His lament that India has adopted a 'victim mentality' when it really needs to adopt a 'winner mentality' has become less of a concern as over time, India has shown commitment to stay on the bandwagon of globalization. Having realized that globalization is a necessary but not a sufficient condition for high growth production, India has undertaken economic reforms, both internal and external. However, it must be ensured that these reforms are synchronized so that the pace of both reforms is set right in order to work hand in hand to promote agricultural productivity growth.

Thus, training the farmers and educating them appropriately to change their mindset and reorienting them to take up new activities or adopt foreign technology is of utmost importance. In this context, it is necessary to involve non-governmental organizations in training and mobilizing the rural poor to face the challenge of liberalization.

Also, with domestic economic reforms, more care needs to be exercised to draw up state-specific liberalization measures to maximize their benefits. Lastly, in the implementation of these reforms for successful globalization, one crucial element, not entirely within control is the need for good governance and stability in the political and economic environment. Political leaders who are the ultimate decision makers in these matters need to examine their own role dispassionately.

It is quite apparent that at this relatively early stage, there is little observable evidence of gains to India's agricultural performance after opening up. However, there could easily be benefits that have not yet surfaced, or are yet to be identified and perhaps too difficult or intangible to measure.

Whatever the case, it is highly likely that it is too soon to assess the full impact of globalization and economic reforms. Furthermore, the process of liberalization has been gradual and remains incomplete. For example, the complete removal of quantitative restrictions after March 2001 will have provided an opportunity for Indian farmers to tap world markets and, if they are successful, results should start to become evident soon. Export promotion via the development of export and trading houses as well as effective liberalizing export promotion zone schemes for agriculture are fairly recent measures and only time will tell as to how effective these measures are. Other possibilities such as agro-industry parks for promoting exports are also in the pipeline.

In conclusion, India has successfully set sail on the waters of globalization and economic reforms and even in the wake of economic and political instability, she has to carefully steer her course in order to reap the benefits of increased productivity growth in the agricultural sector.

ACCELERATING AGRICULTURAL GROWTH

Poverty in India is predominantly a rural phenomenon. About 70% of the population, and about 85% of the poor, live in rural areas and most depend on agriculture. Agriculture provides livelihood to 60 percent of the rural people and remains vital for food security.

KEY ISSUES AND CONSTRAINTS

Deceleration in agricultural growth. In the last decade (1995/96-2004/05) the agricultural GDP growth rate slowed to less than 2% per year, compared to about 3.5% per annum in the preceding decade. In the poorest states, such as Madhya Pradesh, Orissa, and Rajasthan, growth in the last decade was below 1% per year. The stagnation of agriculture, the high proportion of poor dependent on it, and the widening rural-urban income gaps are major concerns of the Government of India (GOI).

Low Agricultural Productivity. With fixed land and water availability, higher agricultural growth can be achieved only by increasing productivity per unit of these resources through effective use of improved technology. Per hectare yields of major crops (foodgrains, oilseeds, other cash crops) in India are lower than many other major producing countries and well below those obtained in crop trials within India.

Moreover, crop yields in many of the poorest states (Bihar, Orissa, Rajasthan, North-east states) are far below those of richer states (Punjab, Haryana, Tamil Nadu). The virtual collapse of the agricultural extension system in most states limits farmer access to improved technologies and practices. Weaknesses in the agricultural research system, such as an irrigated agriculture bias, weak prioritization and proliferation of programmes, top-down supply driven programmes, weak cost effectiveness, undermine the effectiveness of the public agricultural research system. Natural resources degradation.

Sustainability of the land and water resource base is at risk with increasing soil degradation and overexploitation of groundwater in many areas. In addition to erosion, salinity and alkalinity, soils are also losing carbon and micronutrients due to unbalanced fertilizer use. Nearly 30% of the blocks in the country are presently classified as semi-critical, critical or overexploited as groundwater use exceeds the rate of groundwater recharge.

Composition of public expenditures. Public expenditures on agricultural subsidies including for power, irrigation and fertilizers, have crowded out productivity-enhancing investments, such as for irrigation development, agricultural research and extension, rural roads and electrification. According to the Planning Commission, budgetary subsidies in agriculture increased from around 3% of agriculture GDP around the late-1970s to about 7% in the early 2000s. During the same period public investment in agriculture declined from 3.4% of agriculture GDP to 1.9%.

GOVERNMENT OF INDIA'S RESPONSE

The Government of India's (GOI) goal is to raise agriculture's growth rate to 4% per year in the 11th Plan. To achieve this, the Plan aims to:

- Double the growth rate of irrigated areas and improve water management in rainfed areas
- Reclaim degraded land
- Bridge the knowledge gap through effective extension
- Foster diversification to higher value agricultural products
- Promote animal husbandry and fisheries
- Facilitate access to credit at affordable rates
- Improve the incentive structure for markets.

Water Resource Management and Irrigation. Water is a critical input for achieving higher agricultural growth and ensuring greater food security. Presently about 40% of cultivated area in India is irrigated. GOI is putting emphasis on more effective use of existing irrigation potential, expansion of irrigated area, and better water management in rainfed areas.

The GOI Accelerated Irrigation Benefits Program aims to assist states to finish the construction of uncompleted major and medium irrigation schemes. Many states, however, lack the policy, regulatory, and institutional framework for efficient, sustainable, and equitable allocation and use of water, or for capturing the environmental costs of inefficient use.

Many states often do not allocate sufficient public funds for operations and maintenance of canals, which lead to the rapid deterioration of irrigation canals and reduce the availability of water to farmers. Limited cost recovery also limits funds for operations and maintenance and undercuts farmer incentives to use water more efficiently, leading to waterlogging and salinity problems in some areas. Some states have adopted participatory irrigation management on a wide scale to improve the management and sustainability of surface irrigation systems.

Free power to farmers by some states or highly subsidized power encourages the excessive use of ground water. This has led to the increased proportion of over-exploited areas in the country and large fiscal costs to state governments. The political economy of power sector reform however impedes progress in this area.

Rainfed areas, where large numbers of poor live, cover 60% of cultivated area in India. Watershed management, rainwater harvesting and ground water recharge can help augment water availability in rainfed areas.

The GOI recently established the National Rainfed Areas Authority to develop an action plan for rainfed areas. In moving forward, some challenges that would need to be addressed include: (i) multiplicity of programmes and agencies (MoA, MoRD, MoEF, donors) and lack of consistency among programmes within the state (scope, community participation, contributions); (ii) weak coordination among various agencies; and (iii) weak local institutions to ensure effective implementation.

Bridging the Knowledge Gap. The GOI is providing additional resources for the National Strategic Research Fund. However, it would be critical to improve the governance and implementation structure of the Indian Council of Agricultural Research system to improve its research effectiveness.

The Ministry of Agriculture (MoA) is seeking to improve extension effectiveness by scaling up the Agricultural Technology Management Agency (ATMA) approach which builds on a bottom-up approach to extension planning and implementation.

It would be important to build on the factors of success derived from the ATMA pilots which include decentralized decision making and funds flow, priority to institutional capacity building and training of extension staff on the new bottom-up approach to extension planning and implementation and ensuring convergence of funds from the various departments at the district level.

The GOI is also encouraging states to develop district plans for agricultural development based on agro-ecological potential. These district plans will be the basis for the formulation of the state agricultural development plans which would be supported under the proposed additional central assistance of Rs. 250 billion announced by the Prime Minister in May 2007. Ensuring adequate stakeholder consultation and community participation in the development of these plans would strongly influence the success of this programme.

Diversification of Production and Income Sources. Rising incomes are driving the increased demand for higher-value fresh and processed agricultural products in domestic markets and globally.

This opens new markets and income opportunities for farmers. Many of these commodities, however, are highly perishable and inadequate rural infrastructure (markets, rural roads, electrification) and services (market information, risk management) increase losses and inhibit the more rapid expansion of high value production. The GOI recently launched the National

Horticulture Mission — the single largest agriculture plan scheme — to promote increased horticulture production.

The Bharat Nirman Program, which supports the expansion of irrigation, rural roads, electrification and telecommunication) will help reduce losses and marketing cost and improve the competitiveness of local products.

The MoA is also encouraging states to amend their Agricultural Produce Marketing Committee (APMC) Acts to liberalize the marketing of agriculture produce at the state level and improve farmer access to markets. So far 11 states have adopted the amendments. In addition, MoA is also promoting public-private partnerships to develop markets and is providing investment grants to encourage investments in agro-processing and marketing.

Continuing trade and domestic regulatory reforms are encouraging increased investments in rural areas. These include the removal of most domestic marketing controls—storage, movement, credit, wholesale marketing, and food processing. However, reforms of the institutional and policy framework for land are needed.

Many states have computerized their land records/registration, but reforms of the institutional and data structures are essential to reduce the cost of land transfers and enhance tenure security in rural and urban areas (for example, eliminate overlapping property claims and facilitate access to credit). The bans or restrictions on land leasing limits access to land by poor and landless rural households and drives tenancy underground. They also limit the productivity of land use.

Diversification of incomes through rural non-farm activities is an important element of the overall strategy to address rural poverty. More rapid growth of the rural non-farm sector is, however, constrained by government interventions in labor, land, and credit markets and inadequate and poor quality infrastructure and services. To improve access to credit in rural areas the GOI is supporting reforms of the credit cooperative system.

Ensuring Food Security. While pursuing crop diversification, an emerging concern is the stagnation in foodgrain production which is raising food security concerns. The 11th Plan approach paper notes that to achieve the goal of a 4% overall agricultural growth rate per year, foodgrain yields per hectare need to increase by at least 2% to 3% per year to compensate for the possible shift in cultivated area from foodgrains to high value crops. To enable this, a concerted effort is needed to raise the relatively low foodgrain yield levels in Eastern and Central India in particular.

Improving Public Expenditures. In recent years the GOI and state government have exerted greater effort to cap the amount of the various subsidies, while increasing expenditures on more productivity enhancing investments. The latter includes several centrally sponsored programmes, such as the National Horticulture Mission, Bharat Nirman rural infrastructure programme, and the National Rural Employment Guarantee Scheme.

Participation of the poorest states in these programmes is hampered, however, by lack of capacity to develop appropriate plans and implement them effectively.

THE ROLE OF GLOBALIZATION AND ECONOMIC REFORM

The Indian agricultural sector has been undergoing economic reforms since the early 1990s in the move to liberalize the economy to benefit from globalization. This paper traces this process, analyses its effects on agricultural productivity and growth and discusses the problems and prospects for globalization to draw policy implications for the future of Indian agriculture.

India, which is one of the largest agricultural-based economies, remained closed until the early 1990s. By 1991, there was growing awareness that the inward-looking import substitution and overvalued exchange rate policy coupled with various domestic policies pursued during the past four decades, limited entrepreneurial decision making in many areas and resulted in a high cost domestic industrial structure that was out of line with world prices. Hence the new economic policy of 1991 stressed both external sector reforms in the exchange rate, trade and foreign investment policies, and internal reforms in areas such as industrial policy, price and distribution controls, and fiscal restructuring in the financial and public sectors. In addition, India's membership and commitment to World Trade Organization (WTO) in 1995 was a clear sign of India's intention to take advantage of globalization and face the challenge of accelerating its economic growth.

One measure of economic growth is given by productivity growth as it forms the basis for improvements in real incomes and welfare. The concept of productivity growth gained importance for sustaining output growth over the long run as input growth alone is insufficient to generate output growth because of diminishing returns to input use. This paper, which examines India's productivity growth in the agricultural sector in the context of globalization, has three main aims. First, it examines these possible links in the agricultural sector in general. Second, it discusses the problems and prospects for agricultural productivity growth of various Indian states. Third, the paper highlights the challenges of globalization and draws policy implications for the success of Indian agriculture.

OVERVIEW OF INDIA'S AGRICULTURAL ECONOMY

In the early 1950s, half of India's GDP came from the agricultural sector. By 1995, that contribution was halved again to about 25 per cent. As would be expected of virtually all countries in the process of development, India's agricultural sector's share has declined consistently over time as seen in the table below.

Table : Share of agricultural output in India's GDP

Year	*1950/51*	*1965*	*1976*	*1985*	*1991*	*1999*
Percentage share	52.2	43.6	37.4	32.8	28.3	24.4

Source: Estimated from various issues of Economic Survey, Government of India.

In the last five decades, the Government's objectives in agricultural policy and the instruments used to realize the objectives have changed from time to time, depending on both internal and external factors. Agricultural policies at the sectoral level can be further divided into supply side and demand side policies. The former include those relating to land reform and land use, development and diffusion of new technologies, public investment in irrigation and rural infrastructure and agricultural price supports. The demand side policies on the other hand, include state interventions in agricultural markets as well as operation of public distribution systems. Such policies also have macro effects in terms of their impact on government budgets. Macro level policies include policies to strengthen agricultural and non-agricultural sector linkages and industrial policies that affect input supplies to agriculture and the supply of agricultural materials.

During the pre-green revolution period, from independence to 1964-1965, the agricultural sector grew at annual average of 2.7 per cent. This period saw a major policy thrust towards land reform and the development of irrigation. With the green revolution period from the mid-1960s to 1991, the agricultural sector grew at 3.2 per cent during 1965-1966 to 1975-1976, and at 3.1 per cent during 1976-1977 to 1991-1992. Acharya (1998) explains that the policy package for this period was substantial and consisted of:

(a) introduction of high-yielding varieties of wheat and rice by strengthening agricultural research and extension services,
(b) measures to increase the supply of agricultural inputs such as chemical fertilizers and pesticides,
(c) expansion of major and minor irrigation facilities,
(d) announcement of minimum support prices for major crops, government procurement of cereals for building buffer stocks and to meet public distribution needs, and
(e) the provision of agricultural credit on a priority basis. This period also witnessed a number of market intervention measures by the central and state Governments.

The promotional measures relate to the development and regulation of primary markets in the nature of physical and institutional infrastructure at the first contact point for farmers to sell their surplus products.

Acharya (1998) also notes that the rate of growth of productivity per hectare of all crops taken together increased from 2.07 per cent in the decade ending 1985-1986 to 2.51 per cent per annum during the decade ending 1994-1995. Similar evidence of an increase in yields, a partial measure of productivity gains given by output per unit of land area is seen below for various crops.

Although productivity gains were sustained in the 1990s after the liberalization process began, the yield rates for most of the agricultural products in India are far below comparable rates in a number of other countries. Except for sugarcane, tea, coffee and jute, India's yields are lower than the world average. It should be noted that India is ranked second both in area and output for sugarcane production and is the largest producer of tea and jute in the world. Although India is doing quite well in wheat production, the average yields in the Netherlands and Ireland are more than three times India's yield rates. In all other major crops, India's productivity performance seems to lag behind others.

Table : Yield for various crops (kg/ha)

	1950/51	*1960/61*	*1970/71*	*1980/81*	*1990/91*	*1995/96*	*1998/99*
Rice	668	1 013	1 123	1 336	1 740	1 855	1 905
Wheat	663	851	1 307	1 630	2 281	2 483	2 596
Coarse cereals	408	528	665	695	900	941	1 035
Pulses	441	539	524	473	578	552	661
Food grains	522	710	872	1 023	1 380	1 499	1 611
Oil seeds	481	507	579	532	771	851	948
Cotton	88	125	106	152	225	246	240
Sugarcane	33 422	45 549	48 322	57 844	65 395	68 369	69 288

Source: Agricultural Statistics At A Glance. Agriculture, CMIE, September 1999.

Why Globalize?

Globalization in the context of agriculture can be best discussed in the context of three components – improvement of productive efficiency by ensuring the convergence of potential and realized output, increase in agricultural exports and value added activities using agricultural produce, and finally, improved access to domestic and international markets that are either tightly regulated or are overly protected.

These components are linked in various ways. For example, productive efficiency would enhance value added activities in agriculture through agro-processing and exports of agricultural and agro-based products. These activities in turn would increase income and employment in the industrial processing sector. Thus globalizing agriculture has the potential to transform subsistence agriculture to commercialized agriculture and to improve the living conditions of the rural community.

Ahluwalia (1996) explains that this indirectly requires an improvement in agricultural growth from between 2 and 3 per cent in the past to about 4 per cent per year. Although initially, with respect to agriculture, there was no major policy reform package in the 1990s, it was however anticipated that the opening up of the agricultural sector to foreign trade, the move to a market determined exchange rate and reduction of protection for industry would, over time, benefit the agricultural sector.

Manmohan Singh (1995), the then Finance Minister, in his inaugural address at the 54th Annual Conference of the Indian Society of Agricultural Economics, brought to notice that a policy of heavy protection of the industrial sector operated to the disadvantage of the agricultural sector when industrial prices were raised relative to world prices and thus the profitability of investing in industry was raised relative to agriculture. This would lead to a shift of resources from agriculture to industry. A policy of heavy industrial protection also led to an appreciation of the exchange rate. Ahluwalia (1996) noted that over-valuation of the exchange rate (before the Indian rupee was devalued by 18 per cent in two phases starting in July 1991) discouraged agricultural exports more than industrial exports because Indian industrial policy had sought to offset the constraints faced by industries via a system of export incentives for market support. Agricultural exports on the other hand were denied any such incentives as they did not use imported inputs.

Ahluwalia (1996) argued that in the past, the agricultural sector was negatively protected because of the above two reasons and the fact that farmers were denied access to the world markets due to trade barriers. Exports of plantation crops and a few commercial crops were free from export restriction but exports of essential commodities, particularly food products, were subject to bans, quotas and other restrictions. Interestingly, Kruger and others (1991) showed that while many developed countries continue to protect agriculture, developing countries do not do so. However, no formal attempt or theoretical framework has yet been used to assess the extent of negative protection in Indian agriculture. The implementation of economic reform in the Indian agricultural sector has been a gradual process. These include an 87 per cent cut in tariff on agricultural products, sustenance of high-yield crop varieties, removal of minimum export price on selected agricultural products, a lift on quantity restrictions on the export of some crops and various land reforms related to tenancy rights and land ceilings.

PRODUCTIVITY GAINS FROM GLOBALIZATION AND ECONOMIC REFORMS

In the wake of India's efforts towards globalization and economic reforms, the expected benefits of total factor productivity (TFP) growth can be represented using the production frontier. The production frontier traces out the maximum output obtainable from the use of inputs.

Opportunities from globalization and economic reforms can lead to:

- shift from A to B due to technical efficiency
- shift from B to C on existing frontier due to input growth
- upward shift from C to D due to technological progress

Which constitute various sources of TFP growth, can be linked with trade gains. The movement from A to B led by technical efficiency allows increases in output when inputs and technology are used to their fullest potential to obtain the greatest yield. Given that India has been involved in agricultural production for so long, there would be learning-by-doing gains that can help boost production given the expected increase in demand as India opens up. The increased production would enable a better utilization of inputs, especially that of advanced capital technology. The reduction in the tariff rate for agricultural products from 113 per cent in 1990-1991 to 26 per cent in 1997-1998 is also expected to motivate local producers into rethinking their production techniques and efficiently utilizing the inputs and technology to keep costs of production down in order to remain competitive. The optimum or efficient use of land and water resources would then allow agriculture to respond to the demand for other products such as horticulture and livestock which is expected to increase following a rising trend in the per capita incomes of both rural and urban groups.

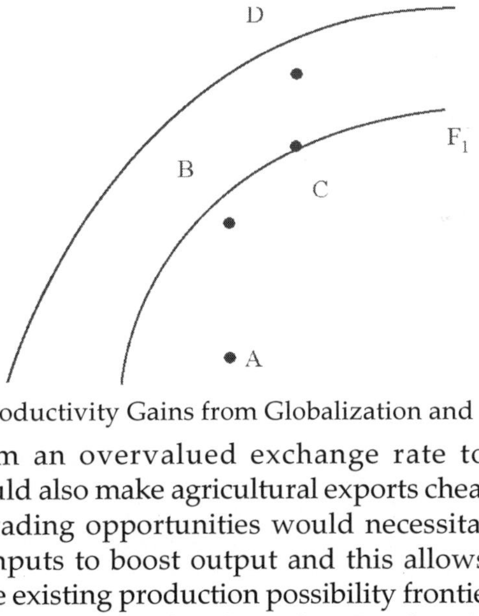

Fig. Total Factor Productivity Gains from Globalization and Economic Reforms

The move from an overvalued exchange rate to that of a market determined rate would also make agricultural exports cheaper and hence boost exports. The new trading opportunities would necessitate an increased use in the quantity of inputs to boost output and this allows for the movement from B to C along the existing production possibility frontier. Increased exports would bring about economies of scale and as Verdoon's law states, output growth would lead to productivity growth.

The scale of output under increased exports would justify the huge fixed costs underlying technologically advanced equipment and hence increase incentives to adopt high quality inputs. The use of such inputs would result in technological progress and this is represented by the shift from C to D. The reduction in tariff rates in industry from 1990-1991 to 1997-1998 range from 153 per cent to 25 per cent for consumer goods, 77 per cent to 18 per cent for intermediate goods and 97 per cent to 24 per cent for capital goods.

This means that farmers now have relatively cheaper access to imported new technology and better capital equipment as well as the option of adopting better farming techniques and this should lead to technological progress. In particular, the development of agro-processing as an instrument for agricultural and rural modernization will bring benefits, given its capital-intensive and technology-intensive nature.

Lower duty rates on plastics and metals also lower costs of packaging. These forms of cost efficiency should allow competitive pricing of products. In addition, external competition can be expected to motivate local producers into the production of improved quality intermediate inputs for agriculture.

The importance of technology in agricultural development was first demonstrated in the 1970s with impressive growth in yields following the introduction of new wheat and rice varieties. But this technology was limited to areas of assured irrigation as the new seeds also required heavy inputs of fertilizers and pesticides for optimal results. However, the potential for further extending this technology is not yet exhausted as there is scope for expanding irrigation further and improving the quality of irrigation in many areas. For further technological progress, genetic engineering and the biotechnology revolution provides a prospect of developing new varieties that can flourish with less dependence on water and chemical inputs. Such reduced dependence upon chemical fertilizers and pesticides is also desirable because of environmental considerations, which are an increasing concern.

It must however be acknowledged that the link between trade liberalization and productivity growth is two-way as they both feed on each other. The productivity gains can be obtained from openness but to benefit from openness via increased demand for exports, agricultural products need to be priced competitively. In other words, productivity growth is necessary to lower the costs of production.

IMPACT OF GLOBALISATION ON THE AGRICULTURAL SECTORS

Agriculture is the engine of most African economies and in recent years governments have become convinced that, by liberalising their economies, agriculture would prosper and provide the necessary growth to provide investment to improve the country's infrastructure, to form the foundation for industrialisation and to improve public services. In the 1980s some African countries began to reform their economic policies. Internal conflict, however, has delayed reform in many ECA countries and they have only recently begun to liberalise their internal economies. Over the last twenty years there has also been an accelerating trend to liberalise trade on a global scale. The success of these global reforms and internal liberalisation measures (for those countries that have adopted them) has been patchy. Most African countries have found it difficult to compete with more efficient foreign agricultural producers and are suffering surges in imported products which compete with domestic production. At the same time, the expected improvement in exports of these products has not materialised. This may be due, in part, to the difficulties of complying with the high quality standards required by many importing countries. The international market prices of almost all agricultural commodities have fallen to their lowest levels (in real terms) in living memory. This is due to over-production encouraged by the export-orientated economic policies of competing producing countries. The cut in agricultural subsidies in the developed world has reduced surplus stocks of food which has had the effect of reducing supplies available for food aid.

The lowering of tariff barriers by consuming countries has offered more opportunities to exporters in developing countries. The exposure of agricultural production to foreign competition has forced some producers to become more efficient. The dismantling of government-controlled marketing boards has stimulated the evolution of the private sector trading networks needed in a modern economy. Some actors in the agricultural sector have seen little benefit from the liberalisation process, however.

Communities of small-scale, isolated farmers (which make up the majority of the population in many ECA countries) find it more difficult to obtain inputs and credit. Extension services have been significantly reduced and the value of their surplus production has fallen. They are especially vulnerable to changes in production systems. The trend towards larger farms and plantations in the name of efficiency has marginalised many rural groups thus adding to the problem of unemployment, urbanisation and cultural disintegration. The effort required to address these problems and challenges will have to be made by many agencies. Trade agreement negotiators, government agencies, agricultural development and agricultural research organisations, NGOs and private-sector farmers', traders' and processors associations all need to be fully aware of the changing relationships between markets and all the different actors in agriculture, and shape their policies and programmes accordingly.

This report examines the main benefits and disadvantages of the trend towards globalisation and the liberalisation of agricultural markets. It also attempts to offer some suggestions on how ECA countries might maximise the opportunities offered by a more open global trading system and how they might respond to some of the negative aspects of liberalisation. In particular, the report will highlight key issues that need to be addressed by all agencies involved in agricultural development in an increasingly globalised economic environment.

EARLY TRADE AND COLONIALISM

East Africa has a tradition of trade, especially with the Arabian peninsular and Southern Asia, going back several thousand years. Arab merchants developed trading links with many African kingdoms and established settlements on the coast before the 8th century AD. Trade routes were established into the interior of the continent to transport commodities such as ivory, gold, furs, gums and slaves, to the coast and Asian food products such as bananas and coconuts as well as some spices were introduced.

European colonisation began in the 15th Century. The Portuguese charted the coastline and developed natural harbours for use in their trade with the East and they too began trading in African goods. The Portuguese were followed by the French and British and later by the Germans and Italians. Each colonial power began to explore further into the interior and to subdue the indigenous populations. The British, especially, recognised the potential for agriculture in the region and established plantations and farms. Large areas were colonised and territorial borders delineated. Cotton was produced as well as food crops and livestock. Local African agriculture was transformed with the introduction of crops from other parts of the world, notably those originating in the Americas such as sweet potatoes, potatoes, maize, cassava, peppers, tomatoes, papaya, jackfruit, cocoa, passion fruit, pineapple, sisal, cashew nuts, sunflower, groundnuts and tobacco. In the late 19th century the first railways were driven deep into central Africa. In certain instances workers were transported to Africa, especially from the Indian sub-continent, to work in the plantations and as labourers to build roads and railways. Indians with an entrepreneurial background were able to establish businesses that grew to represent a significant proportion of trading activity in some countries.

Europe's main commercial interest in ECA was its raw materials – minerals and agricultural products. The present transport system reflects those interests. Railways and major roads were designed to carry products from the interior to the coast, not to encourage or facilitate trade within the region. Borders between countries are largely arbitrary, from an African point of view, and they often cut across the territory of populations with a common language and culture.

European companies dominated import/export trade in African products often holding a monopoly in specific commodities. Many of these companies not only organised the production of these commodities but also transported them and processed them locally or, more often, in their home country. Some, like the giant chocolate, tea, sugar, tobacco and coffee companies, also marketed the finished product. In order to protect these monopolies trade barriers had to be erected between African countries colonised by different colonial powers. At times, countries of the region were officially at war with each other as the French fought the English, the English fought the Germans, and so on. Such an environment, of course, discouraged the evolution of trading links within the region. As urban populations grew and the number of plantation workers increased, the Europeans found it necessary to organise the distribution of food through governmental structures and to control both purchasing and retail prices.

PROCESS OF DE-COLONIALISATION

The process of de-colonialisation began after World War II and most African countries gained their independence in the 1950's and 1960s. African governments generally retained the marketing structures and trade barriers bequeathed to them by the colonial regime but they benefited from the relatively high commodity prices at the time. In the 1970s and early 80s some commodities, such as coffee, cocoa, sugar and rubber, were the subject of International Commodity Agreements which maintained prices at a level agreed by producing and consuming countries. Some African countries experimented with collectivised production as a means of improving the economies of scale and others concentrated on self-sufficiency and import substitution. Many large commercial farms and plantations were retained and expanded.

In the post World War II period, European countries embarked on programmes to develop their own agricultural sectors, partly in response to their experiences in the war, when large quantities of food had to be imported by sea at great cost both in money and lives. All developed countries were able to boost agricultural output through innovations in farming technology – machinery, artificial fertilisers, pesticides and new multiplication techniques. Farms too became bigger and more efficiently managed. During this period, developed countries also enjoyed unprecedented rates of industrial and economic growth. Trade in manufactured goods and services increased enormously. Transport systems improved by air, sea and land and communications systems were developed to facilitate trade. Trade in raw materials and agricultural products has become a relatively minor component of international commerce. Farming now represents only about two percent of total economic output in the most highly developed countries. Traditional African farming methods do not lend themselves so easily to efficient, large-scale production.

Conflict

Burundi, Eritrea, Ethiopia, Democratic Republic of the Congo, Rwanda, Sudan and Uganda have all suffered multiple problems from war or internal conflict. Their economies have been severely damaged by military expenditure, degradation of infrastructure, dislocation of populations, loss of labour to military activities and neglect of, and damage to, farms and factories. Millions of ECA citizens have lost their lives over the last thirty years. Apart from the devastation of war itself, the disruption caused by conflict has delayed the implementation of economic reforms and development programmes and many of these countries have lost their access to aid programmes and investment opportunities.

The reasons for conflict have included tribal animosity, territorial disputes, remnants of Cold War conflict, totalitarianism and religious intolerance. Some analysts have identified a link between conflict and poverty in which it is difficult to break the cycle of poverty leading to unrest and unrest exacerbating poverty. At the end of 2001, however, there seem to be some grounds for optimism. Most major conflicts in the region have been resolved or, at least reduced to a lower level of intensity.

AGRICULTURAL DEVELOPMENT STRATEGIES

The lack of development in African countries has caused their economies to fall further and further behind those of the leading industrial nations. Many different development strategies have been tried. Some African countries have successfully encouraged investment in mining, tourism and industry. In agriculture, producers have been encouraged to move away from subsistence farming towards a more commercial approach as governments realised that income generated from the sale of surplus production could be used to improve productivity.

Agricultural development in ECA has faced an uphill struggle for the last twenty years. In an effort to stimulate development many countries borrowed heavily from bodies such as the IMF and from the commercial banking sector. These loans were not granted without strings attached, however. Most African countries were obliged to liberalise their economies by adopting significant policy changes often applied in packages known as Structural Adjustment Programmes (SAPs). These programmes included a number of elements but generally included requirement to:

- devalue the currency (to discourage imports and make exports more competitive),
- to make the currency freely convertible with other currencies,
- to cut public expenditure (in order to lower taxes),
- to dismantle state controlled marketing boards,
- to privatise state-owned industries (to raise capital and stimulate competition),
- to cut import restrictions (to encourage local industries to become more efficient),
- to allow foreign companies to freely repatriate profits (to encourage inward investment),
- and to boost exports.

The economists who designed SAPs were convinced that the only way African countries could transform their economies was to encourage inward investment and earn foreign exchange to invest in infrastructure and lay the foundations for industrialisation. These measures assumed that any country could compete in the world market if production and investment was concentrated in areas where they were deemed to have a competitive advantage. The only activity in which ECA nations could be said to have a competitive advantage in the world market was in the production of agricultural products and the exploitation of natural resources such as forestry, fishing and mining. The major flaw in this strategy was that similar advice was given to almost all tropical countries at the same time. Coffee-producing countries were encouraged to boost coffee production; sugar producers should produce more sugar, and so on. This resulted in over-production of these commodities which caused prices to plunge in the international markets. On average, current prices of tropical products (taking dollar inflation into account) are only about one seventh of those prevailing in 1980 (UN General Assembly). Economists call this phenomenon the fallacy of composition - less income is earned as more commodities are produced.

Another component of SAPs which many observers believe to have been counter-productive was the requirement to cut public expenditure. All too often this meant a cut in health programmes, education and agricultural extension. These measures have tended to reduce, rather than enhance the flexibility of the workforce and to curtail agricultural development. Overall, the record of inward investment has been poor and the ending of currency controls has increased opportunities for transfer pricing abuse (where companies over-price imports and under-price exports to reduce tax liability).

The most important SAP reform affecting the distribution of agricultural products has been the dismantling of state-controlled marketing boards and the practice of setting fixed purchasing and sales prices for commodities. It was assumed that government control of markets had obscured the forces of competition in supply and demand in the economy. A free market system would unleash these forces and increase productivity. It would force producers to meet the demands of consumers both in price and quality. Farmers would be able to buy inputs cheaper from competing suppliers, and the country, as a whole, would become more competitive in world markets.

Unfortunately, competitive and transparent markets did not emerge spontaneously (Shepherd). Most African farmers have too little land to produce truck-loads of goods and they are widely dispersed over the countryside. There is not enough business to encourage more than one trader to operate in many areas. Farmers have no means of communicating with the outside world or even the nearest town and they are often unwilling to risk the investment of bringing their goods to market resulting in considerable waste. Laws may have been passed which ban collusion among traders to pay low prices to farmers and charge high prices to consumers, but there are often insufficient resources to enforce such laws. Most traders have no experience of free market conditions and are reluctant to put their fellow traders out of business with serious competition.

Advocates of SAPs point to examples of countries that have improved their economies after adopting SAPs (World Bank) but there are few in Africa. Most ECA countries were not able to implement SAPs until relatively recently but rates of poverty have increased in many of these countries. Intense conflict, both within and between countries of the region, drought, desertification and, now HIV/AIDS have further weakened economic development in ECA. Most critics of the reform process, however, acknowledge that markets in African countries must be made more competitive and SAPs are designed to do that but this process may take a considerable time.

Trade Agreements

Economic links between ECA countries and their former colonial rulers have been maintained since independence. The economies of these countries have been moulded to meet the needs of their European counterpart for a hundred years or more and it would have been difficult for them to make the necessary changes in production patterns to trade successfully with other countries. The Europeans too needed to maintain supplies of raw materials and export markets in Africa and to protect the business of their trading companies.

In 1975 all ten countries covered by this study became party to the Lomé Convention. The Convention established trade, aid and cultural relationships between 15 European countries and 71 so called ACP (African, Caribbean and Pacific) countries which had either been colonies of, or had had strong historical links with, Europe. This agreement did not rule out bilateral or multilateral agreements with other countries but did give ACP countries preferential access to European markets. ECA countries have also decided to try to stimulate regional trade by bringing their economies closer together in regional economic agreements such as COMESA and SADC.

East Africans have exchanged goods and ideas with many other peoples of the world for millennia. In these exchanges of goods, cultural links have been established which have influenced ECA life at all levels – in religion, the arts, public sector structures, the economy and agriculture. In the last decade or two, however, this process has accelerated tremendously. There is no agreed definition of globalisation. It is simply a term which has been used recently to describe the impact of innovations in communication and transport systems on trade and the growing interdependence of nations due to economic sophistication and burgeoning output. In addition, high levels of protection between trading blocks of countries are breaking down as barriers to trade are reduced. These changes have made it possible to increase the volume of trade between countries in agricultural products.

It became clear that overall levels of trade could be increased if trade barriers were reduced, where there was agreement to do so, and that international trade should be governed by mutually agreed rules. The most active trading nations have been keen to find new markets for their goods and to reduce the barriers to free trade. These countries, however, have been reluctant to expose their own markets to foreign competition, especially unfair competition from subsidised or sub-standard goods.

At the international level, global liberalisation was stimulated by the General Agreement on Tariffs and Trade (GATT) which was first implemented in 1948 as a mechanism to promote free and fair trade among member countries. Several rounds of negotiations of trade rules have occurred throughout the history of GATT. The Uruguay Round, which began in 1986, was the eighth of the GATT rounds. In April 1994, officials from more than 100 countries gathered in Marrakech, Morocco to sign the Uruguay Agreement and to confer the role of further trade reforms on the newly established World Trade Organisation. The reform process is by no means complete. Almost all countries have now committed themselves to the objectives associated with their membership of the WTO. (6 out of the 10 countries covered in this study are WTO members). In order to meet these objectives, countries are obliged to further reform their existing internal economic and external trade policies. The future of trade and agriculture in ECA is inextricably linked to the rate and direction of these reforms.

4

Poverty Alleviation in Rural India : Programmes and Strategy

Poverty eradication is one of the major objectives of planned development. The magnitude of the problem is still quite staggering. Thirty six per cent of the Indian population was below poverty line (BPL) in 1993-94, the latest year for which the data are available and the absolute number of poor was 320 million, out of which 244 million (37 per cent of the rural population) lived in rural areas. The incidence of poverty declined from 54.9 per cent in 1973-74 to 36 per cent in 1993-94. But the absolute number of poor did not decline much over this period of 20 years. There were 321 million poor in 1973-74 and 320 million in 1993-94; in the rural areas the corresponding numbers were 261 million and 244 million.

The main determinants of poverty are (i) lack of income and purchasing power attributable to lack of productive employment and considerable underemployment and not to lack of employment per se; (ii) a continuous increase in the price of food, especially foodgrains which account for 70-80 per cent of the consumption basket; and (iii) inadequacy of social infrastructure, affecting the quality of life of the people and their employability.

Economic growth is important. Economic growth creates more resources and has the potential of creating more space for the involvement of the poor. But the involvement of the poor depends on the sources of growth and the nature of growth. If the growth is sourced upon those sectors of the economy or those activities which have a natural tendency to involve the poor in their expansion, such growth helps poverty eradication. Therefore, it is important to source a large part of economic growth in agriculture, in rural non-agricultural activities and in productive expansion of the informal sector which all have high employment elasticities, as well as in an export strategy based on labour intensive exports.

The Government recognises that high growth of incomes is by itself not enough to improve the quality of life of the poor. Unless all the citizens of the country, and most particularly the poor, have certain basic minimum services, their living conditions cannot improve. These minimum services include among other things literacy education, primary health care, safe drinking water and nutritional security. The Government had convened a meeting of Chief Ministers to identify such basic minimum services and a list of seven services had unanimously been agreed upon. These seven services are safe drinking water, primary health facilities, universal primary education, nutrition to school and pre-school children, shelter for the poor, road connectivity for all villages and habitations, and the Public Distribution System (PDS) with a focus on the poor. The Ninth Plan lays special emphasis on these seven basic minimum services and will make all efforts to achieve a minimum level of satisfaction in providing these in partnership with the State Governments and the Panchayati Raj Institutions (PRIs).

Direct poverty alleviation programmes are important and will continue on an expanded scale in the Ninth Plan. But these programmes would be oriented towards strengthening the productive potential of the economy and providing more opportunities for involving the poor in the economic process. Broadly, there would be schemes for income generation through supplementary employment, for the welfare of the poor in rural/urban areas and for a targeted PDS system to ensure that the poor have access to foodgrains at prices they can afford. In this chapter, both rural and urban poverty alleviation programmes besides the Targeted Public Distribution System (TPDS) will be discussed in some detail. Poverty can effectively be eradicated only when the poor start contributing to the growth by their active involvement in the growth process. Implementation of the programmes should be increasingly based on approaches and methods which involve the poor themselves in the process of poverty eradication and economic growth. This is possible through a process of social mobilisation, encouraging participatory approaches and institutions and empowerment of the poor. In this the PRIs, the voluntary organisations and community based Self-Help Groups will be more closely involved.

RURAL POVERTY ALLEVIATION

SELF-EMPLOYMENT PROGRAMMES

The Integrated Rural Development Programme (IRDP) aims at providing self-employment to the rural poor through acquisition of productive assets or appropriate skills which would generate additional income on a sustained basis to enable them to cross the poverty line. Assistance is provided in the form of subsidy and bank credit. The target group consists largely of small and marginal farmers, agricultural labourers and rural artisans living below the poverty line. The pattern of subsidy is 25 per cent for small farmers, 33-1/3 per cent for marginal farmers, agricultural labourers and rural artisans and 50 per cent for Scheduled Castes/Scheduled Tribes families and physically handicapped persons. The ceiling for subsidy is Rs.6000/-for Scheduled Castes/ Scheduled Tribes families and the physically handicapped; for others, it is Rs.4000/-in non-DPAP/non-DDP areas and Rs.5000/-in DPAP and DDP areas. Within the target group, there is an assured coverage of 50 per cent for Scheduled Castes/Scheduled Tribes, 40 per cent for women and 3 per cent for the physically handicapped. Priority in assistance is also given to the families belonging to the assignees of ceiling surplus land, Green Card Holders covered under the Family Welfare Programme and freed bonded labourers.

IRDP is a Centrally Sponsored Scheme which is in operation in all the blocks of the country since 1980. Under this scheme Central funds are allocated to States on the basis of proportion of rural poor in a State to the total rural poor in the country. Since the inception of the programme till 1996-97, 50.99 million families have been covered under IRDP at an expenditure of Rs.11434.27 crore. The total investment during this period has been Rs.28047.65 crore which includes a subsidy component of Rs.9669.97 crore and a credit disbursement of Rs.18377.68 crore. Of the total families assisted under this programme 44.75 per cent were Scheduled Castes/Scheduled Tribes and 27.07 per cent women.

During the Eighth Five Year Plan the total allocation (Centre and State) under IRDP was Rs.5048.29 crore and the total investment amounted to Rs.11541.06 crore. In quantitative numbers, 10.82 million families were covered under IRDP against the initial target of 12.6 million families fixed for the entire Eighth Plan period. However, from 1995-96 physical targeting under the programme was abolished with the focus shifting to financial targets and qualitative parameters. Of the families covered 50.06 per cent were Scheduled Castes/Scheduled Tribes and 33.59 per cent women. The coverage of women was still lower than the target of 40 per cent.

The IRDP has been successful in providing incremental income to the poor families, but in most cases the incremental income has not been adequate to enable the beneficiaries to cross the poverty line on a sustained basis mainly because of a low per family investment. The results of the last Concurrent Evaluation (September 1992-August 1993) revealed that of the total beneficiaries assisted under the programme, 15.96 per cent of the old beneficiary families could cross the revised poverty line of Rs.11,000 (at 1991-92 prices), while 54.4 per cent of the families were able to cross the old poverty line of Rs.6,400 per annum. But, the analysis by income group of families revealed that in case of those within initial income of Rs.8501 – 11,000, 48.22% of beneficiary families could cross the poverty line of Rs.11,000 which is quite encouraging. The analysis of the family income of the beneficiaries reveal that a large percentage (57.34%) of the families had annual family income from assets of more than Rs.2000. The annual income from the asset was more than Rs.6000 in 29% cases.

The major constraint in the implementation of IRDP has been sub-critical investments which have adversely affected the Incremental Capital Output Ratio (ICOR) levels and thereby undermined the viability of the projects. Though the average per family investment has been rising steadily in monetary terms, in real terms the increase has been inadequate and in some cases sub-critical due to the inflationary trends and the increase in the cost of assets.

At the instance of the Ministry of Rural Development (now renamed as Ministry of Rural Areas & Employment), the Reserve Bank of India appointed in 1993, a High Powered Committee under the Chairmanship of Dr. D.R. Mehta, Deputy Governor of Reserve Bank of India to make an indepth study of IRDP and recommend suitable measures for its improvement. The Committee was asked to review among other factors, the process of selection of appropriate income generating assets, credit structure, recovery of loans, and procedural matters in respect of obtaining loans, and efficacy of existing administrative structures of the District Rural Development Agencies (DRDAs). In consonance with the recommendations of the High Powered Committee, the new initiatives taken by Government under IRDP in the Eighth Plan included (a) targeting the segment of literate unemployed youth below the poverty line for IRDP activities by giving them subsidy upto Rs.7500 or 50 per cent of the project cost (whichever is lower) (b) promotion of group activities through enhancement of ceiling on subsidy to Rs.1.25 lakh or 50 per cent of the project cost (whichever is lower) for all group ventures involving at least 5 members (c) back-ending of subsidy to prevent leakages in subsidy administration (d) shifting the emphasis to financial targets and qualitative parameters from a perfunctory physical coverage of families and (e) enhancing the limit of allocation to programme infrastructure from 10 per cent to 20 per cent in all the States and 25 per cent in the North Eastern States.

Among the other steps taken to enhance the efficacy of the programme are abolition of the cut of line to enable all families below the poverty line to be assisted under the programme, targeting the investment per family at progressively higher levels each year, extension of the family credit plan to 213 districts of the country, enhancing the ceiling limit of collateral free loans to a uniform limit of Rs.50,000 with a view to easing the constraints faced by poor beneficiaries while taking loans from the banks, extension of the cash disbursement scheme to 50 per cent blocks in the country, decentralisation of the sanctioning powers for infrastructural projects below Rs.25 lakh and entrusting the banks with the task of identification of beneficiaries in about 50 districts on a pilot basis. These interventions have had an impact on the average per family investment which rose from Rs.7889 in 1992-93 to Rs.15036 in 1996-97.

In pursuance of the High Powered Committee's recommendation, for the first time in 1995-96 credit targets were fixed. There has been a continuous increase in the volume of credit mobilised by the banks during the successive years of the Eighth Plan period. Correspondingly, the subsidy credit ratio, which averaged 1:1.77 in the first three years of the Eighth Plan, rose to 1:1.96 in the fourth year and further to 1:2.17 in 1996-97. However, there are genuine reasons for the inability of the banks to meet the full credit requirements of IRDP beneficiaries. These include poor recovery of IRDP loans, lack of adequate rural banking infrastructure in certain areas and the weak financial performance of Regional Rural Banks and Cooperative banks.

There has been considerable diversification of IRDP activities since the inception of the programme. Initially, a majority of the beneficiaries under the programme subscribed to primary sector activities. In 1980-81 the sectoral composition of IRDP activities was heavily skewed towards the primary sector which had a sponsorship of 93.56 per cent, while the share of the secondary and tertiary sectors were 2.32 per cent and 4.12 per cent respectively. Over the years, the share of the primary sector has come down considerably and is currently around 55 per cent, while the shares of the secondary and tertiary sectors have increased proportionately to 15 per cent and 30 per cent respectively.

Inadequate development of infrastructure and insufficient forward and backward linkages and market facilities have been an area of concern under IRDP. In an attempt at filling up the critical infrastructural gaps and strengthening the linkages and marketing facilities, the allocation under IRDP towards the development of programme infrastructure was increased from 10 per cent to 20 per cent in all the States and to 25 per cent in the North Eastern States. Decentralisation in the sanctioning powers for infrastructural projects had already been given effect to in 1994-95. However, despite this enhanced provision for programme infrastructure under IRDP and the relaxation in sanctioning norms, the actual expenditure on infrastructural development was a mere 5 per cent to 7 per cent of the total allocation under the programme at the all-India level. There is, therefore, a critical need to prepare a perspective infrastructural plan at the district and block level and to ensure that the funds earmarked for infrastructural development under IRDP are closely monitored and not diverted elsewhere.

TRAINING OF RURAL YOUTH FOR SELF-EMPLOYMENT (TRYSEM)

The Scheme of TRYSEM, a facilitating component of IRDP, aims at providing basic technical and entrepreneurial skills to the rural poor in the age group of 18 to 35 years to enable them to take up income generating activities. The Eighth Plan had emphasised the importance of a proper assessment of the training needs of the rural youth in relation to self and wage-employment opportunities, quality of training and group training. During the Eighth Plan, 15.28 lakh youth were trained under TRYSEM, of whom 34.16

per cent took up self-employment and 15.05 per cent wage-employment; while the remaining 50.79 per cent remained unemployed.

With a view to strengthening this programme, several initiatives were taken in the Eighth Plan which include, among others, an increase in the stipend and honorarium rates; emphasis on professionalised training through the established and recognised institutes like ITIs, Community Polytechnics, Krishi Vigyan Kendras etc., exploring the possibilities of setting up production groups from amongst TRYSEM trainees for undertaking ancillary activities like manufacture and assembly of modern items of production; utilisation of TRYSEM infrastructure funds for the strengthening of Nirmithi Kendras (Rural Building Centres) sponsored by HUDCO for training of youth under TRYSEM in the trades of low cost housing and the setting up of mini-ITIs at the block level to strengthen the training infrastructure for the rural youth.

The TRYSEM programme was evaluated for the first time in a Quick Study (June to August 1993) conducted through independent research institutes/organisations. The main findings of the evaluation study are as under :

- Of the total sample districts, area skill surveys were not carried out in 92 per cent of the districts to assess the potential skill requirements. This resulted in a mismatch of job skills in 53.3 per cent of the sample districts.
- Of the total number of beneficiaries, who got training under TRYSEM, roughly 47.19 per cent were unemployed after the training and 32.54 per cent took up self-employment after training of whom 12.41 per cent took up employment in trades other than those in which they were trained.
- A majority of the beneficiaries *i.e.* 66.52 per cent cited lack of funds as a major reason for not taking up self-employment independently after the training.
- A major proportion of TRYSEM trainees *i.e.* 53.57 per cent did not apply for loan under IRDP. Of the total beneficiaries, who applied for loan, only about 50 per cent were given assistance under IRDP upon completion of training.
- Roughly, 73.38 per cent of the beneficiaries could derive an average monthly turnover upto Rs.1000 as a result of self-employment taken up by them after the training.
- 63 per cent beneficiaries felt no improvement in their socio-economic conditions as a result of TRYSEM training.

There has been a poor convergence of TRYSEM with IRDP which has also been reflected in the Fourth Round of the Concurrent Evaluation of IRDP (1992-93). Only 3.88 per cent of the IRDP beneficiaries had received training under TRYSEM. It was also observed that the rural youth trained under TRYSEM were only interested in the stipendiary benefits they received during the course of training and therefore, had not utilised the knowledge gained under the programme for furthering their self-employment prospects. In practice, therefore, such expenditure on training had become infructuous because of an absence of linkages between the employment opportunities available and training provided. Clearly, TRYSEM has been a weak link in the overall strategy for self-employment in rural areas.

SUPPLY OF IMPROVED TOOLKITS TO RURAL ARTISANS (SITRA)

Launched in July 1992, as a sub-scheme of IRDP in selected districts, this scheme has since been extended to all the districts of the country. Under the scheme, a variety of crafts persons, except weavers, tailors, needle workers and beedi workers, are supplied with a kit of improved hand tools within a financial ceiling of Rs.2000, of which the artisans have to pay 10 per cent and the remaining 90 per cent is a subsidy from the Government of India. The supply of power driven tools, subject to a ceiling of Rs.4500, is also permitted under this scheme. Beyond this, any additional finance required by the artisans can be provided through loans under IRDP. The rural artisans are trained under TRYSEM for which an age relaxation has been provided to them.

Since the inception of this scheme in 1992-93 upto 1996-97, 6.10 lakh toolkits have been distributed to rural artisans at an expenditure of Rs.116.19 crore. Reports from the State Governments indicate that the scheme has been well received by rural artisans.

The more popular crafts under this scheme are blacksmithy, carpentry, stone craft, leather work, pottery and cane & bamboo work. Prototypes of improved tools in these crafts have been developed by the National Small Industries Corporation (NSIC), Regional Design and Technical Development Centres under the Development Commissioner, Handicrafts and other organisations.

The SITRA was evaluated by an independent research organisation, *i.e.* Development Alternatives, New Delhi, in two Districts of Uttar Pradesh, namely Agra and Aligarh. The findings of this study reaffirms the positive impact of SITRA. It also indicates that the income level of rural artisans have increased substantially with the use of improved tools.

DEVELOPMENT OF WOMEN AND CHILDREN IN RURAL AREAS

The special scheme for Development of Women and Children in Rural Areas (DWCRA) aims at strengthening the gender component of IRDP. It was started in the year 1982-83, on a pilot basis, in 50 districts and has now been extended to all the districts of the country.

DWCRA is directed at improving the living conditions of women and, thereby, of children through the provision of opportunities for self-employment and access to basic social services. The main strategy adopted under this programme is to facilitate access for poor women to employment, skill upgradation, training, credit and other support services so that the DWCRA women as a group can take up income generating activities for supplementing their incomes. It seeks to encourage collective action in the form of group activities which are known to work better and are more sustainable than the individual effort. It encourages the habit of thrift and credit among poor rural women to make them self-reliant. The programme also envisages that this target group would be the focus for convergence of other services like family welfare, health care, nutrition, education, child care, safe drinking water, sanitation and shelter to improve the welfare and quality of life of the family and the community.

Since the inception of the scheme till 1996-97, 1,87,918 DWCRA groups were formed at an expenditure of Rs.248.95 crore, covering 30,39,383 rural women. It was in the Eighth Plan that DWCRA received a fillip with the Government taking several initiatives to strengthen the programme. These include, among others, extending its coverage to all the districts of the country, increasing the revolving fund from Rs.15,000 to Rs.25,000, permitting the formation of smaller DWCRA groups in difficult terrain and remote areas, and permitting operation of joint accounts by the group organiser and another member of the group elected as treasurer of the group rather than the Gram Sevikas and the group organiser, so as to facilitate the DWCRA groups in managing their own affairs. The Child Care Activities (CCA) component was introduced in the DWCRA programme in 1995-96 with the objective of providing child care services for the children of DWCRA women. Similarly the Information, Education and Communication (IEC) component was introduced to generate an awareness among rural women about the development programmes being implemented for their upliftment and welfare. The Eighth Plan also saw the extension of the Community Based Convergent Services (CBCS), a component of DWCRA, to 141 districts of the country.

DWCRA – THE CASE OF ANDHRA PRADESH

- Strategy adopted for formation of sustainable DWCRA groups had the following salient features :
- Formation of Thrift and Credit groups to develop group dynamics, cohesion and homogeneity among the members.
- Savings provided the entry point for poor women to come together through a Self-Help mechanism.
- Democratically managed groups with collective decision making.
- Sustainable income generating activities with access to credit under the Integrated Rural Development Programme (IRDP) and to training facilities.

- Total Literacy Campaign (TLC), Kalajatha, multimedia publicity campaign through All India Radio (AIR). Doordarshan and print media, involvement of youth leaders, mahila mandals, voluntary organisations and Government functionaries created awareness and contributed to the process of social mobilisation

In the implementation of DWCRA, some States like Andhra Pradesh, Kerala, Tripura and Gujarat have performed very well while in other States, the performance and impact of DWCRA has been relatively poor. In Andhra Pradesh, in particular, several successful DWCRA groups have been formed and this has led to the empowerment of women in decision making on various social aspects that impinge on their daily life. The range of activities pursued by these groups are also fairly diverse. Some have started mini banks and have, thereby, reduced their dependence on the money lenders. Other groups are managing lands taken on lease. Quite a few have formed mini transport companies, having acquired autos, LCVs etc. on bank loans. The success of this programme has been attributed to two major catalysts namely, adult literacy among women and its culmination into a women's movement and close involvement of the NGOs. There is a need to evolve an institutional mechanism for replicating the successful DWCRA groups throughout the country.

Pride of Nellore Women'-a Success Story

- As a sequel to the Total Literacy Campaign & anti-arrack agitation, 'PODUPULAKSHMI' (Podupu means saving, Lakshmi – Goddess of Wealth) was started by two lakh women organised into 7000 small thrift groups of 20-30 members. They saved Rs.8.00 crore in four years.
- The district administration provided the 'PODUPULAKSHMI' movement with the initial support and acted as a facilitator. Volunteers, animators, trainers and instructors in the Total Literacy Campaign acted as PODUPULAKSHMI organisers.
- Once PODUPULAKSHMI groups reached a level of maturity, they converted themselves into DWCRA groups. This enabled them to access the revolving fund under DWCRA which was used by the groups to provide working capital to set up micro enterprises. The savings fund was used to meet emergency consumption needs.
- Today a wide variety of women centred activities are carried out by these groups. The ANM, the School Teacher, the Fair Price Shop Dealer, the Anganwadi Worker are all associated with PODUPULAKSHMI – bringing about a convergence of basic services

Yet, in the implementation of DWCRA several shortcomings have also surfaced which has stymied its successful and effective execution in some States. Several groups have become defunct over time. The reasons for these include, among others, (a) improper selection of groups; (b) lack of homogeneity among the group members; (c) selection of non-viable economic activities which are mostly traditional and yield low income; (d) the linkages for supply of raw material and marketing of production are either deficient or not properly planned as a result of which DWCRA groups have become vulnerable to competition. The District Supply and Marketing Societies have been weak outlets for the sale of DWCRA products; (e) lack of institutional financial support, inadequate training, a non-professional approach and poor access to upgraded technological inputs have deprived DWCRA groups from diversifying into high value addition activities; and (f) inadequacy of staff and their insufficient training and motivation has also affected the overall implementation of the programme. These shortcomings would have to be suitably addressed for the successful implementation of the programme in the Ninth Plan.

WAGE EMPLOYMENT PROGRAMMES

JAWAHAR ROZGAR YOJANA (JRY)

Rural poverty is inextricably linked with low productivity and unemployment. Hence, it is imperative to improve productivity and increase employment in rural areas. An employment-oriented growth strategy would achieve this goal only in the medium and long run. In the short run supplementary employment will have to be provided to the unemployed and underemployed, during the lean agricultural season. There are two major wage employment programmes namely the Jawahar Rozgar Yojana (JRY) and the Employment Assurance Scheme (EAS) presently in operation.

The JRY was launched as a Centrally Sponsored Schemes (CSS) on 1st April, 1989 by merging the National Rural Employment Programme (NREP) and the Rural Landless Employment Guarantee Programme (RLEGP). The main objective of the programme is the generation of additional gainful employment for unemployed and underemployed persons, both men and women, in the rural areas through the creation of rural economic infrastructure, community and social assets with the aim of improving the quality of life of the rural poor.

The resources under this scheme are allocated to the States/UTs on the basis of proportion of rural poor in the States/UTs to the total rural poor in the country. From the States to the districts, the allocation is made on an index of backwardness which is based on the proportion of rural Scheduled Castes/ Scheduled Tribes population in the district to total Scheduled Castes/ Scheduled Tribes population in the State and an inverse of agricultural production per agricultural worker in that district, in equal weights. The funds are devolved to the village panchayats by giving due weightage to the Scheduled Castes/Scheduled Tribes population and the total population of the village panchayat. Until recently, these funds were distributed between the village panchayats and the district level in the ratio of 80:20. However, subsequent to the revitalisation of PRIs at three levels, the JRY funds are now distributed among the village panchayats, intermediate panchayats and the district panchayats in the ratio of 70:15:15.

This programme is targeted at people living below the poverty line. However, preference is given to Scheduled Castes/Scheduled Tribes and freed bonded labourers. Atleast 30 per cent of the employment is to be provided to women under the Yojana. In practice, however, this programme is self targeting. Given that employment is offered at statutory minimum wages for unskilled labour and that these wage rates are generally lower than the prevailing market wage rates, only those willing to do manual work for the prescribed wage rates would seek employment on these public works. While works under the scheme can be taken up during any part of the year whenever the need for generating employment is felt, these should preferably be started during the lean agricultural season but may continue thereafter, if necessary.

After three years of its implementation, *i.e.* in 1992-93, a review of the programme revealed that the per person employment generated was inadequate in terms of the requirement and did not provide enough income to the poor. It was also perceived that the resources under JRY were too thinly spread and adequate attention was not being given to the backward areas of the country. Accordingly, the strategy for implementation of JRY was modified from 1993-94 with the introduction of the Second Stream of JRY, specifically targeted at 120 identified backward districts in 12 States of the country, characterised by a concentration of the poor and the underemployed, with additional resources flowing to these districts. This modification in programme strategy was made to achieve the target of providing 90-100 days of employment per person in backward districts where there was a concentration of unemployed and underemployed persons. In addition, a Third Stream of JRY was introduced for taking up special and innovative projects aimed at preventing migration of labour, enhancing women's employment and undertaking special programmes through voluntary organisations for drought proofing etc.

A Concurrent Evaluation of the JRY was conducted from June 1993 to May 1994. The study revealed that nearly 82.16 per cent of the available funds were spent on community development projects. Construction of rural link roads received the highest priority. The wage and non-wage component of the expenditure of JRY works undertaken by the village panchayats was of the order of 53:47 at the all-India level against the stipulated norm of 60:40. Muster rolls were maintained with 86.87 per cent of the village panchayats. The average wages paid per manday of the unskilled workers were more or less on the lines of the minimum wages stipulated under the Act. Of the assets created, 76.96 per cent were created by the village panchayats and 76.11 per cent of these assets were found to be in a good condition. As many as 69.35 per cent of the workers were satisfied with the benefits they received under the JRY.

The Evaluation Report also brought into focus certain inadequacies in the programme. It was reported that 57.44 per cent of the elected panchayat heads had not been imparted any training for the implementation of JRY works. The share of women in employment generated under the programme was only 16.59 per cent and 49.47 per cent of the works could not be completed on time on account of shortage of funds. Other shortcomings observed were differentials in the wages paid to male and female workers, non-utilisation of locally available material in a large number of JRY works undertaken by panchayats and lack of discussion of the annual action plans in the Gram Sabha meetings etc.

In a comprehensive restructuring of the wage employment programmes on 1.1.1996, JRY was further streamlined. In the revised strategy, the First Stream of JRY was continued in its existing form but Indira Awaas Yojana (IAY) and Million Wells Scheme (MWS) which were till then sub-schemes of JRY were made independent schemes. The Second Stream of JRY, which was being implemented in 120 backward districts in the country, was merged with the Employment Assurance Scheme (EAS) introduced in 1775 selected backward blocks of the country in 1993-94 in view of the similarity in these programmes. The Third Stream of JRY with its thrust on innovative projects was continued. Accordingly, the JRY is now being implemented in two parts *i.e.* (i) the Jawahar Rozgar Yojana (Main Scheme); and (ii) Special and Innovative Project. Since the inception of JRY in 1989-90 till 1996-97 a total amount of Rs.26570.25 crore (Centre and State) was allocated to the programme. As against this total allocation, an amount of Rs.25661.70 crore was released of which Rs.25190.30 crore was utilised by the States. This utilisation is approximately 98.16 per cent of the total funds released. In terms of physical performance, as against a target of 6581 million mandays fixed under JRY, the actual employment generated was 6585 million mandays which is an achievement of 100.07 per cent. Of the total employment generated under the programme, the share of Scheduled Castes/Scheduled Tribes was 3659.53 million mandays (55.57 per cent) and that of women 1681.40 million mandays, which is 25.53 per cent.

From the data on mandays of employment generated under JRY it is difficult to assess the number of workers who have actually received employment in the rural areas and on an average for how many days. To surmount this shortcoming, registration of workers who take up employment under this programme, should be made compulsory as under the EAS.

Besides generating supplementary employment of a casual manual nature, the programme has contributed to the development of rural infrastructure through the creation of a wide range of community and social assets in a number of sectors. These included major irrigation works, soil conservation works, land development, drinking water wells, rural roads, construction of school buildings, panchayat ghars, mahila mandals, houses and sanitary latrines and social forestry. In fact, assistance for construction of class rooms under the Operation Black Board (OBB) programme was specially provided under JRY. Of the total cost, 60 per cent was funded from JRY and 40 per cent from the education department of the State Governments to meet the additional non-wage cost. Under Operation Black Board, 25576 classrooms and 21541 school buildings were constructed at an expenditure of Rs.176.11 crore from JRY funds between 1991-92 to 1994-95. Such an integration between sectoral programmes and JRY, with dovetailing of funds, would help in the creation of better quality durable assets.

EMPLOYMENT ASSURANCE SCHEME (EAS)

The Employment Assurance Scheme was launched on 2nd October, 1993 in 1775 identified backward blocks situated in drought prone, desert, tribal and hill areas, in which the revamped public distribution system was in operation. Subsequently, the scheme was extended to additional blocks which included the newly identified Drought Prone Area Programme (DPAP)/Desert Development Programme (DDP) blocks, Modified Area Development Approach (MADA) blocks having a larger concentration of tribals, and blocks in flood prone areas of Uttar Pradesh, Bihar, Assam and Jammu & Kashmir. In addition, 722 non-EAS blocks previously covered under Second Stream of Jawahar Rozgar Yojana (JRY) were also brought under the EAS. The EAS has since been universalised to cover all the rural blocks in the country with effect from 1.4.1997.

The main objective of the EAS is to provide about 100 days of assured casual manual employment during the lean agricultural season, at statutory minimum wages, to all persons above the age of 18 years and below 60 years who need and seek employment on economically productive and labour intensive social and community works. The works are to be selected by the District Collector and implemented through the line departments in such a manner that the ratio of wage to the non-wage component would stand at 60:40. Sectoral norms for execution of various works are-watershed development (50 per cent) and agro-horticulture, minor irrigation works (10 per cent) in DPAP and DDP blocks or water & soil conservation including afforestation, agro-horticulture and silvipasture (40 per cent), and minor irrigation works (20 per cent) in non-DPAP/non-DDP blocks. In addition, funds are also earmarked for link roads featuring in the Master Plans developed in the respective districts for this purpose (20 per cent) and public community buildings in rural areas as per the felt needs of the districts (20 per cent). The village panchayats are involved in the registration of persons seeking employment and the panchayats maintain these registers. They also coordinate and monitor the works. A maximum of two adults per family are to be provided employment under the scheme. The applicants, who register themselves for employment under the EAS, are issued family cards in which the number of days of employment are entered as and when such employment is given to them.

The EAS is a Centrally Sponsored Scheme. The scheme is demand-driven and therefore no fixed allocations are made for the districts/blocks. Instead, initial notional allocations are made to districts at the commencement of each year and thereafter depending on the demand for supplementary employment and the actual utilisation of funds the districts can request for additional funds. For the purpose of initial release, the blocks have been classified into three categories *i.e.* category A, B & C and Central funds to the tune of Rs.40 lakh, Rs.30 lakh and Rs.20 lakh are released as the first of the two instalments to these blocks respectively. This corresponds to the notional minimum allocation of Rs.1 crore, Rs.75 lakh and Rs.50 lakh per block per annum including the State's matching share.

Since the inception of EAS in 1993-94 (*i.e.* October 1993) upto 1996-97, a total amount (Centre and State) of Rs.6514.65 crore has been released under the programme, against which the total utilisation was Rs.5278.16 crore. This indicates a percentage utilisation of 81.02 per cent. As many as 25.90 million persons registered themselves for employment under the EAS. The programme generated 1068.60 million mandays of employment from 1993-94 (October 1993) to 1996-97.

The EAS has not been evaluated till date. Hence, it is difficult to assess the overall impact of the programme in terms of employment, enhancement in the purchasing power of the poor and creation of durable assets. However, the Programme Evaluation Organisation has recently undertaken a comprehensive evaluation study of this Scheme.

MILLION WELLS SCHEME (MWS)

In India, though the small and marginal farmers, with holdings of less than 2 hectares, account for about 78 per cent of the total operational holdings, they only cultivate about 32.2 per cent of the cropped area (Agricultural Census 1990-91). To increase the productivity of these holdings they must be ensured an assured source of water supply. Ground water made available through wells is an important source specially in the remote areas of the countryside, where canal or tank irrigation is not feasible. Though the fixed capital investment in well irrigation is fairly high, it has many advantages such as flexibility in operation, dependability of source, timing of water deliveries and low conveyance losses.

The Million Wells Scheme (MWS) was launched as a sub-scheme of the National Rural Employment Programme (NREP) and the Rural Landless Employment Guarantee Programme (RLEGP) during the year 1988-89. After the merger of the two programmes in April 1989 into the Jawahar Rozgar Yojana (JRY), the MWS continued as a sub-scheme of JRY till December 1995. The MWS was delinked from JRY and made into an independent scheme with effect from 1.1.1996.

The scheme was primarily intended to provide open irrigation wells, free of cost, to individual, poor, small and marginal farmers belonging to Scheduled Castes/Scheduled Tribes and freed bonded labourers with a 20 per cent earmarking of JRY funds. Tubewells and borewells are not permitted under the Scheme. Where wells are not feasible due to geological factors, other minor irrigation works can be undertaken such as irrigation tanks, water harvesting structures as also development of land belonging to small and marginal farmers. From the year 1993-94 the scope of the MWS has been enlarged to cover non-Scheduled Castes/non-Scheduled Tribes small and marginal farmers who are below the poverty line and are listed in the IRDP register of the village. The sectoral earmarking which was 20 per cent upto 1992-93 had also been raised to 30 per cent from 1993-94 with the stipulation that the benefits to non-Scheduled Castes/Scheduled Tribes would not exceed one third of the total funds utilised during the year.

The MWS is also a Centrally Sponsored Scheme. The cost/area norms in regard to works under MWS are decided upon by a Committee comprising of Chief Secretary, Secretary (RD), Secretary (Planning), Secretary (Irrigation) and Chief Engineer (Minor Irrigation) of the State. The beneficiarficiaries themselves are asked to undertake construction of their wells through their own labour and local labour for which they are paid. Contractors are banned under this programme. The wage to material ratio is required to be maintained at 60:40. Supplementary material costs, if any, can be met from other private/ public sources. Though lifting devices are not provided under the scheme, the beneficiaries who intend to install a lifting device, are given the preference under IRDP and other relevant programmes.

The MWS is being implemented throughout the country. Allocations are made to the States/UTs on the basis of the proportion of rural poor in the State/UTs to the total rural poor in the country. The District-wise allocations are made by the States from their allocation in relation to the unirrigated land held by the target group with a potential for well irrigation.

A total of 11.04 lakh wells have been constructed since the inception of the programme till 1996-97 at an expenditure of Rs.4003.11 crore. There has been no evaluation or impact study conducted in the field for the MWS. Yet, on the basis of the feedback available from certain parts of the country, this programme achieved considerable success in the districts falling in the Chotanagpur region of South Bihar, large parts of Orissa, many districts of Gujarat, besides the Eastern and Southern region of Rajasthan. In these areas, the MWS has played a significant role in transforming single cropped dry land areas held by farmers of the target group into double cropped lands, leading to increase in agricultural output and incomes. Yet, such successes have not been uniformly reported across the country. Many States have expressed difficulty in the implementation of the programme. For instance, in Punjab and Haryana where the incidence of tubewell irrigation is widespread and there is a wide network of canal irrigation systems, the programme of open dug wells is a non-viable option. Similarly, in Kerala the small size of the land holdings of the small and marginal farmers gives the scheme a limited potential. Andhra Pradesh, Goa, Madhya Pradesh, Maharashtra, and West Bengal are some of the other States which have shown a poor performance in the construction of wells under the MWS. These States have been permitted to utilise the allocations made under MWS for other schemes of minor irrigation such as irrigation tanks, water harvesting structures and also for the development of land belonging to the small and marginal farmers. Some States have also been permitted to divert MWS resources for the construction of houses for the poor under the Indira Awaas Yojana.

Field studies in various parts of the country have identified several factors which have posed as impediments to the effective implementation of this scheme. These include, among others, (a) construction of wells without proper hydro-geological surveys; (b) a declining water table and its continuous depletion by overuse of pumping sets resulting in large tracts falling in the dark/grey zones which indicate already dangerous levels of depletion of ground water; (c) non-availability of eligible persons in the target group; (d) limited success in rocky and sandy strata; and (e) distance between wells affecting the rate of discharge.

The programmc has also been hamstrung by inadequate linkages. There has been a failure on the part of the block officers and the banks in providing lifting devices under IRDP and other programmes, thus rendering the investment in open dug wells infructuous. In some cases, though the wells have been dug and are working, their full potential has not been realised because of a lack of extension support from the agricultural department.

NATIONAL SOCIAL ASSISTANCE PROGRAMME (NSAP)

The National Social Assistance Programme (NSAP) came into effect from 15th August, 1995. The programme represents a significant step towards the fulfilment of the Directive Principles in Articles 41 and 42 of the Constitution through the enunciation of a National Policy for social assistance benefits to poor households in the case of old age, death of the primary breadwinner and maternity. It is a Centrally Sponsored Scheme with 100 per cent Central assistance provided to States/UTs.

This programme has three components : namely (i) National Old Age Pension Scheme (NOAPS); (ii) National Family Benefit Scheme (NFBS); and (iii) National Maternity Benefit Scheme (NMBS) which are targeted at people living below the poverty line. Under the National Old Age Pension Scheme (NOAPS), old age pension of Rs.75 per month is provided to persons of 65 years and above who are destitutes. The National Family Benefit Scheme (NFBS) provides a lump sum family benefit of Rs.10,000 to the bereaved household in case of the death of the primary bread winner irrespective of the cause of death. This scheme is applicable to all the eligible persons in the age group 18 to 64. Under the National Maternity Benefit Scheme (NMBS) there is a provision for payment of Rs.500 per pregnancy to women belonging to poor households for pre-natal and post-natal maternity care upto the first two live births. This benefit is provided to eligible women of 19 years and above.

In providing social assistance benefits to poor households in cases of old age, death of the primary bread winner and maternity, the NSAP supplements the efforts of the State Governments with the objective of ensuring minimum national levels of well-being.

The NSAP provides opportunities for linking social assistance package to schemes for poverty alleviation and provision of basic minimum services. In fact, that old age pension can be linked to medical care and other benefits aimed at the aged beneficiaries. The Integrated Rural Development Programme/Jawahar Rozgar Yojana assistance may be provided in addition to the family benefit for the families of poor households, who suffer the loss of the primary bread winner. Maternity assistance can be linked to other programmes of maternal and child care.

The NSAP is implemented in the States/UTs through Panchayats and Municipalities. The Panchayats and Municipalities are encouraged to involve Voluntary agencies to the extent possible for identifying beneficiaries and persuading them to avail of the benefits intended for them.

The NSAP programme has been operationalised since August 1995. No evaluation of the programme has been conducted either by the Central Government or by any other institution/organisation so far.

In the first year of the programme in 1995-96, as against the total allocation of Rs.538.93 crore, the release was Rs.380.49 crore. Of the total release, only Rs.183.77 crore was actually utilised. Consequently, in 1996-97 there was a large opening balance of Rs.197.53 crore. Again, as against the total allocation of Rs.916.21 crore, the total release in 1996-97 was Rs.548.29 crore. Thus, the total available funds in 1996-97 was Rs.745.82 crore, of which only Rs.383.50 crore was utilised (*i.e.* 51.42 per cent utilisation).

Of these three programmes, the performance under the NOAPS has been relatively good as compared to that of NFBS and NMBS because the administrative machinery for the implementation of this programme was already in place in most of the States. If we analyse the State-wise physical performance for the year 1996-97, even under NOAPS, only 10 States had covered more than 90 per cent numerical ceilings of the target population. Of this, 5 States/UT namely Andhra Pradesh, Haryana, Karnataka, Madhya Pradesh and Pondicherry had more than 100 per cent achievement. The achievement in many States was less than 50 per cent. These States include Goa, Maharashtra, Rajasthan and North Eastern States excluding Assam.

The performance in the case of NFBS and NMBS was particularly poor. Under NFBS, achievement was only 28.85 per cent of the numerical ceiling. In only five States namely, Andhra Pradesh, Goa, Madhya Pradesh, Punjab and Tamil Nadu the coverage was above 50 per cent and in 7 States/UTs the coverage was between the national average of 28.85 per cent and 50 per cent, while in the remaining 20 States/UTs, the coverage was below the national average.

Under NMBS, achievement was only 27.57 per cent of the numerical ceiling. Except in Andhra Pradesh and Tamil Nadu where the coverage was 51.36 per cent and 60.94 per cent respectively, in all other States/UTs the coverage was below 50 per cent. In the case of 7 States/UTs the coverage was between the national average of 27.57 per cent and 50 per cent, while in the remaining 23 States/UTs, the coverage was below the national average.

Steps would be taken to simplify the procedures for the sanctioning and disbursement of benefits under these schemes with a closer involvement of the Gram Panchayats/Municipalities. The assistance may be sanctioned and disbursed in public meetings preferably of Gram Sabha by either Gram Panchayat functionaries or Block functionaries or appropriate level. In the case of urban beneficiaries, elected local self government officials wherever available should be involved in the process of sanctioning the assistance. The disbursement in such cases should also be preferably made in public meetings of neighbourhood/mohalla committees. These schemes would also be given increased publicity to ensure a larger coverage of beneficiaries.

LAND REFORMS

Land Land is considered as an important element of life and is highly valued in most of the world. Land is useful to man in many ways as a source of food, for place to live, for wood, for place to work etc., In India, Before colonial rule the land used to be in the hands of the community as a whole. However during the British Raj this has changed.

Lord Carnwallies has introduced Permanent Land Settlement for Bengal, Bihar and Orissa in 1793. According to this the tax farmers appointed by the british rulers will be coverted as various Land Lords. Under this rule they have to pay fixed commission to East India Company. Thus these intermediary are formed, called as Jagirdars / Jamindar.

Emergence of Tenants Following the Land Settlement Act, 1793, The farmers purchase lands from the Land Lords and hire it for their agricultural use. These people who hired the land are called Tenants. Variations in Tenency

Cash Tenents : They pay a fixed tax for the use and occupy of the land Share – cash Tenants: They pay part of their rent in cash and other part as share of the crop Crop - share Tenants : They pay a share of crops only Croppers : They pay crop of the share. But they are not independent they work under the landlord. Other unspecified tenants:

Ryotwari : This existed in Madras since 1772. And extended to Bombay Presidency. Under this system., every registered holder of the land is considered as proprietor. He paid revenue directly to the government and is having liberty to sell / lease / gift the property to any one. Mahalwari: The land ownership is held as joint ownership with the village body. The land can be cultivated by tenants who can pay cash / kind / share. Jamindari: Under this system the whole village was under one landlord. The persons interested can work in the jamindars land as tenant / labourer based on the agreement with the jamindar. The jamindari system was known to be more exploitive, as the jaminder used to fix / hike the prices of land when ever he wished to do so. Jagirdari: Almost as similar as Jamindari system. The jagirdar is powered to control the unproductive masses of village by engaging them in agrarian activities.

Despite attempts at land reforms over successive Plan periods, the basic character of the agrarian economy has not undergone any structural change. The pattern of land distribution is highly skewed, with a high concentration of land in the hands of a few land owners on the one hand and the growing number of marginal and sub-marginal farmers on the other. Fragmentation of land holdings continues on a large scale and only a few States like Punjab, Haryana, Uttar Pradesh and parts of Maharashtra have been able to successfully undertake a programme of consolidation of holdings.

Agricultural tenancy, which was abolished in most of the States by various enactments in the post-Independence era, continues unabated though it is largely concealed. In the wake of liberalisation, several State Governments have modified their land ceiling laws so as to exempt orchards, fish ponds etc., from the purview of land ceilings. There is also a move to make suitable changes in tenancy regulations to attract private corporate investment in agriculture. Hence, it is necessary to reconsider the issue of land reforms, particularly from the point of view of the poor, as access to land is still a major source of livelihood in rural India. In fact, it has been argued that the need for poverty alleviation programmes has arisen because the land reforms have not been implemented in a systematic way. The experience of several countries in East Asia shows that land reforms, leading to structural equity in the distribution of land, are an essential prerequisite for economic development through agricultural transformation. In addition, the efficiency of land use and land management, and protection of the land rights of the tribals and women have assumed great significance in the context of the changes that are taking place in rural India.

The continued importance of land reforms was recognised in the Eighth Plan, with the abolition of intermediaries, redistribution of ceiling surplus land, tenancy reforms providing security of tenure to tenants and share croppers, consolidation of holdings and updating of land records as the main objectives of the land reform policy. However, only limited success was achieved with respect to these objectives in the Eighth Plan. Given that land reforms is a State subject, the Central Government can only draw the attention of the State Governments to the pressing needs for land reforms, which are central to any strategy of poverty alleviation.

At the end of the Seventh Plan, out of the 72.2 lakh acres of land declared surplus, only 46.5 lakh acres had been distributed. At the end of the Eighth Plan, out of the total 74.94 lakh acres declared surplus, 52.13 lakh acres had been distributed. In other words, during the Eighth Plan only 6-7 lakh acres were redistributed. Further, 12.4 lakh acres were under disputes pending in courts and 19.59 lakh acres were not available for distribution because they were unfit for cultivation or reserved for public purposes or for other miscellaneous reasons. In fact, only 59,000 acres were available for redistribution. Of the Bhoodan land donated, 53 per cent was distributed, accounting for 24.52 lakh acres. In addition, 142.87 lakh acres of wastelands were distributed among 88.5 lakh beneficiaries. But, there is still considerable scope for redistributing Government wastelands, common lands, ceiling surplus land and Bhoodan land.

Similarly, in the area of tenancy reforms very little progress has been made, after the initial abolition of 'zamindari' and the transfer of title to owner-cultivators in the immediate post-Independence period. The successful implementation of tenancy laws has been confined to West Bengal, Karnataka and Kerala. In fact, in the Eighth Plan there was no progress in respect of conferment of rights on tenants and therefore the issue of tenancy reforms is still illusory, but requires tackling.

Consolidation of holdings has taken place in very few States. While 15 States had enacted appropriate legislation, Andhra Pradesh, Tamil Nadu, Kerala, Pondicherry and the North-Eastern States do not have any laws for consolidations of holdings. Several States like Bihar, Maharashtra and Rajasthan have suspended the programme. In fact, only in Uttar Pradesh, 900-1000 villages are being covered annually.

There is evidence of considerable alienation of tribals from their land. As per the latest available estimates, 4.6 lakh cases of tribal land alienation covering 9.2 lakh acres have been registered. Of these 2.7 lakh cases covering 6.3 lakh acres have been disposed of in favour of tribals but physically an estimated 4.7 lakh acres had been restored to them. In other cases, reconciliations are being effected.

It cannot be gainsaid that the essential prerequisite of any land reform measure is the recording of land rights and their updating. In recognition of this, a Centrally Sponsored Scheme for Strengthening of Revenue Administration and Updating of Land Records was introduced during the later half of the Seventh Plan and against an outlay of Rs.20.8 crore, Rs.14 crore were spent. In the Eighth Plan a provision of Rs.175 crore was made against which Rs.98 crore were released and Rs.66.7 crore utilised.

But, given that the funds are shared in the ratio of 50:50 between Centre and States, several States have not been able to provide their share and hence, the utilisation has been low. Several other States have not availed of this scheme at all. In 1995-96, funds were released only to Bihar, Kerala, Maharashtra, Madhya Pradesh, Rajasthan, Tamil Nadu and West Bengal aggregating to only Rs.18.8 crore.

The States have been requested repeatedly to expedite expenditure and to adopt new technologies for survey and settlement operations, preparation of maps etc. Funds for infrastructure development were also sanctioned to meet the training needs of the revenue functionaries. Some States have developed new training institutions, while a few have upgraded the existing ones.

In 1988-89, 8 pilot projects were taken up for computerisation of land records with 100 per cent Central assistance. Since inception, 323 projects have been sanctioned and funds to the tune of Rs.64.43 crore have been released. However, the utilisation was only to the extent of 20 per cent. This shows the tardy progress made in the implementation of this scheme on the ground.

NINTH PLAN STRATEGY

Direct poverty alleviation programmes will continue on an expanded scale in the Ninth Plan. But these programmes would be oriented towards strengthening the productive potential of the economy and providing more opportunities for involving the poor in the economic process.

The Integrated Rural Development Programme (IRDP) would continue to be the major self-employment programme targeted to families living below the poverty line in the rural areas of the country. In the Ninth Plan it would be implemented through an integrated approach under which the existing schemes of Training of Rural Youth for Self-Employment (TRYSEM) and Supply of Improved Toolkits to Rural Artisans (SITRA), Development of Women and Children in Rural Areas (DWCRA) and Ganga Kalyan Yojana (GKY) would be subsumed into the main programme.

To facilitate higher levels of investment under the programme, there would be a strategic shift under IRDP from an individual beneficiary approach to a group and/or cluster approach. As part of the group approach, the focus of IRDP would be on the formation of Self-Help Groups (SHGs) which would be the catalyst for organising the poor. The cluster approach would focus on the identification of a few specified viable activities based on the local resource endowment and occupational skills of the people of that area. The IRDP will also aim at diversifying the investments into high-value-addition sectors and non-traditional activities which have a market potential. The financial institutions would play a more significant and dynamic role by enhancing the credit flows through a continuous line of credit, instead of a one-time loan and would render constructive assistance to the beneficiaries.

This would help them to increase returns on their investments. The IRDP would service the beneficiaries through a package approach, wherein the beneficiary would have access to credit, training as per requirements, upgradation of technology, delivery of essential inputs and marketing tie-ups in an integrated manner. Presently, under the IRDP, training under TRYSEM is provided as an isolated input and there has been little attempt to make a proper assessment of opportunities where skills could be more gainfully utilised. Hence, the new holistic approach would overcome some of the inherent shortcomings which have undermined the success of the programme.

In view of the near complementarity in the ongoing wage employment schemes, Jawahar Rozgar Yojana (JRY) and Employment Assurance Scheme (EAS) will be rationalised. In the revised format JRY will be confined to the creation of rural infrastructure at the village panchayat level in consonance with the felt needs of the community. To the extent that the works undertaken under JRY would be largely labour intensive, supplementary wage employment would be generated in the process of infrastructure creation. The EAS would be the major wage employment programme which would contribute significantly to the provision of the mandated 100 days of casual manual work to those who register for employment under it. As the EAS has been universalised, specific measures would be taken to ensure that the benefits reach the poor and more backward districts of the country.

Land reforms will continue to be an important policy instrument for alleviating rural poverty. Access to land is still a major source of liveliood and its possession enhances the status of people in rural society. A proper implementation of land laws and policies would lead to a restructuring of the agrarian economy in a way conducive to higher rates of agricultural growth but with greater equity in the distribution of gains from it. While the ingredients of the land reform policy would continue to be the same as before, the focus would shift to a few critical areas. All efforts would be made to detect and redistribute the ceiling surplus land and to enforce the ceiling laws stringently.

Given that small and marginal farms are viable, both from the efficiency and equity points of view, it is desirable that the existing ceiling limits are strictly enforced. More importantly, tenancy reforms would have to be taken up especially in States characterised by semi-feudal modes of production. The rights of tenants and sharecroppers need to be recorded and security of tenure provided to them. Leasing in of land should be made permissible within the ceiling limits. The poor should be given access to wastelands and common property resources. The land rights of women must be ensured. This would require amendment of the existing legislations in some States to ensure women's rights with regard to inheritance of both land owned as also under tenancy. Updating of land records would have to be expedited as this is a necessary prerequisite of any effective land reform policy. Since land reforms is a State subject the States would have to be persuaded to take up these measures.

While the implementation of poverty alleviation programmes would be made more effective in the Ninth Plan, the extant non-monetary policies and institutional arrangements which adversely affect the interests of the poor would also be suitably addressed. In this context, the Ninth Plan will identify those laws, policies and procedures which are anti-poor and take the initiative in exploring whether some of these could be suitably modified in favour of the poor.

In the Ninth Plan, the poverty alleviation programmes would be more effectively integrated with area development programmes and the various sectoral programmes within the umbrella of Panchayati Raj Institutions (PRIs). The PRIs will function as effective institutions of local self-government. These would prepare the plans for economic development and social justice through the District Planning Committees and implement them. The State Government would have to devolve administrative and financial power to the PRIs which fall within their purview as per the Eleventh Schedule of the 73rd Constitutional Amendment Act, 1992. In addition, the voluntary organisations (VO) will have to play a more dynamic role in empowering the poor through advocacy, awareness generation and formation of SHGs.

INTEGRATED RURAL DEVELOPMENT PROGRAMME (IRDP)

New Initiatives under IRDP

- IRDP will be a holistic programme covering all aspects of self-employment, namely, organisation of beneficiaries and their capacity building, planning of activity clusters, infrastructure, technology, credit and marketing.
- The existing sub-schemes of TRYSEM, DWCRA, SITRA and GKY to be merged into IRDP.
- Progressive shift from the individual beneficiary approach to the group and/or cluster approach.
- To facilitate group approach SHGs will be formed and steps will be taken to nurture them.
- For cluster approach each district will identify 4 to 5 activity clusters in each block based on local resources and occupational skills of the people. The infrastructure needs for the identified activities will be met in full.

- The Banks will be closely involved in the planning and preparation of projects, identification of activity clusters, infrastructure planning as well as capacity building and choice of activity of the SHGs.
- Promotion of multiple credit rather than one time credit injection.

The IRDP would continue to be the major self-employment programme, targeted towards families living below the poverty line in the rural areas. However, in the Ninth Plan, the focus would be on pursuing an integrated approach under IRDP by subsuming the existing sub-schemes of Training of Rural Youth for Self-Employment (TRYSEM) and Supply of Improved Toolkits to Rural Artisans(SITRA), Development of Women and Children in Rural Areas (DWCRA) and Ganga Kalyan Yojana (GKY) into the main programme. This integration of schemes is necessary to develop the appropriate forward and backward linkages to achieve a synergistic complementarity in the overall implementation of the programme.

Project for Linking Banks with Self-Help Groups

- NABARD initiated a Pilot project to link Banks with Self-Help Groups (SHGs), with the objective of meeting the credit needs of the poor through formal financial institutions.
- Evaluation studies show success with increase in loan volume of SHGs, shift to income generating activities, nearly 100 per cent recovery and reduction in transaction costs.
- 85% of the groups linked with banks were exclusively of women.
- Scheme is being expanded, and will be replicated all over the country.

Furthermore there would be a strategic shift from an individual beneficiary approach to a group and /or cluster approach.

- To facilitate this process Self Help Groups (SHGs) will be formed under IRDP and steps will be taken to nurture these groups to enable them to function effectively as well asto choose their economic activity. Efforts would be made to involve women members in each SHG. Besides, formation of exclusive women groups will also continue as at present under DWCRA. It is proposed that group ventures which involve at least five beneficiaries would be assisted by making available 50 per cent of the project cost subject to a ceiling of Rs.1.25 lakh. This enhanced level of investment would facilitate economies of scale and improve recovery.
- Alternatively, a cluster approach would be preferred wherein a few specified activities are identified for assistance in an area. This would necessitate the formulation of a menu of "activity-based" project profiles in different sectors to suit the local resource endowment and the occupational skills of the local people.

Accordingly, each DRDA would set up four to five activity clusters. Appropriate infrastructure and technology inputs would be built into the project.

- The Family Credit Plan would be extended to all the districts of the country in a phased manner. The feedback from the States suggests that this strategy has met with reasonable success and has raised investment levels.

One of the major constraints in the implementation of IRDP has been sub-critical investments, which have adversely affected Incremental Capital Output Ratios (ICOR) and, thereby, undermined the viability of the projects. Recognising that the level of investment is the most crucial variable in determining the incremental income generated under IRDP, the credit flows and the average level of investment per family for the Ninth Plan would aim at achieving enhanced levels of investment in the range of Rs.25,000-Rs.50,000 depending on the estimate of the poverty line and the poverty gap. These higher levels of investment will give the beneficiary the necessary financial support for diversifying into high-value-addition sectors and non-traditional activities which have a market potential.

In this effort at achieving higher investment levels, the financial institutions would have to play a more significant and dynamic role by enhancing credit flows and rendering constructive assistance to the beneficiary making their investments viable under this programme. Adoption of simplified procedures by financial institutions would facilitate the BPL families in accessing groups loans under IRDP.

To simplify procedures the regimen of subsidy administration would be suitably streamlined in the Ninth Plan. At present, there is a complex gamut of ceiling limits on subsidy. The ceiling limit on subsidy would be fixed at 30% of the project cost subject to a maximum of Rs.7,500. However, in the case of SC/ST it would be fixed at 50% of the project cost subject to a ceiling of Rs.10,000. Therefore, for purposes of administrative expediency the area specific differential in subsidy administration would no longer exist. For group activities, however, the subsidy would continue to be Rs.1.25 lakh or 50 per cent of the project cost, whichever is lower.

The IRDP will seek to develop close linkages with the credit mechanism in such a manner as would promote repeated/multiple doses of credit rather than a one-time loan for the beneficiary. The emphasis, therefore, would be on establishing a continuous line of credit for the beneficiary, wherein it would be possible for the borrower to obtain need-based additional credit for working capital purposes, meeting unforseen expenditure related to proper maintenance of assets etc. which would have a bearing on the viability of the project so as to sustain credible levels of income generation.

Experience has shown that the IRDP has been relatively more successful in land-based activities. In recognition of this fact, purchase of land was made a permissible activity under the programme. For land-based activities, besides providing assistance for purchase of inputs to enhance the productivity of land, there exists a potential for diversifying into other allied activities, which have a high value-addition, such as sericulture, aqua-culture, horticulture and floriculture, on the existing lands of the small and marginal farmers as well as on land leased by the landless. In addition to these activities, SHGs would be given 'pattas' for development of wastelands, social forestry, soil conservation and watershed projects. Common property resources would also be allocated to SHGs in the villages on a long-term basis for eco-sensitive resource management. There is also a potential for allotting nurseries of the forest department to these groups for management. In all these, interest of women's groups would have to be protected. Usufruct rights of women on minor forest produce would have to be ensured legally and administratively and assistance provided to them for purchase or leasing in of land for joint management.

Yet, it must be recognised that land is a limited resource and that the distribution of land holdings remains highly skewed. Hence, the need for exploring the potential of the non-farm sector is crucial. It cannot be gainsaid that the non-farm sector in the rural areas has witnessed both growth and diversification in the past few years. Around 50 per cent of the IRDP investments are now made in the secondary and tertiary sectors, based on local resources and local requirements. These include processing industries, handlooms and handicrafts.

Again, in most villages there is scope for tailoring and ready-made garments, chemist shops, woodcraft, country tiles, general store etc. In addition, in villages of a reasonable size, other business/service ventures like flour milling, motor rewinding, cycle repair etc. could also be promoted under IRDP.

Greater emphasis should be placed on developing rural industries which would catalyse the overall development of that area. In a situation of an ever changing pattern of demand, emergence of new products and technologies in a dynamic and growing economy, an attempt would be made to integrate the IRDP activities, particularly those in the Industry Services and Business (ISB) sector with the market. This task would be entrusted to a team of experts/professionals, who would prepare sectoral and micro plans for identifying the thrust areas and activities for working out necessary linkages with other departments/agencies.

The artisans in the rural areas, despite their rich heritage and skills, largely belong to the poverty group. The scheme for Supply of Improved Toolkits to Rural Artisans is directed to this particular target group. In the Ninth Plan, to enable the rural artisans to take advantage of the new opportunities thrown up by the market, there has to be a quantum jump in their skills and productivity to make their activities viable and profitable. It is also necessary to support them with appropriate product designs and training, improved technology on the one hand and professional management and marketing support on the other. In other words, the rural artisans should be serviced through a package approach wherein the distribution of improved toolkits is supplemented with the supply of credit and raw material, marketing support and upgradation of existing technology.

The timely and adequate supply of trade specific toolkits would be given added attention. While deciding the source for supply of toolkits, quality and cost considerations, alongwith the post-delivery services offered by the manufacturer would be of considerable significance. It is important to constantly upgrade the design of the toolkits. For this, research and development would be given due emphasis. States would be advised to take the initiative in developing toolkits in at least two or three artisanal trades. The Department of Science and Technology, the National Small Industries Corporation, the Small Industries Services Institutes and the IITs would interact with the State Governments for development of appropriate technology and designs for improved toolkits for various artisan groups. In the process of designing and manufacturing these trade-specific toolkits, there would be constant consultations and dialogue with groups of artisans and reputed craftsmen in that particular trade. To broadbase the source for supply of toolkits some master craftsmen would also be trained in the manufacture of improved tools/toolkits. An effort would also be made to develop capabilities for design and upgradation of improved toolkits by artisans themselves. There are pockets of rural technology which have survived the onslaught of modern technological innovations by virtue of their sturdiness and locale-specific utility. These would be identified and replicated elsewhere. For the dissemination of technology the State Governments, in collaboration with the manufacturing agencies, would undertake promotional activities by organising exhibitions at several locations and on important occasions.

Since the 1991 Census data on rural artisans are still not available, the States would be required to conduct a district-wise survey of the total number of rural poor artisans, the number of such artisans provided with improved toolkits and the balance number to be covered. This exercise would facilitate a planned and phased coverage of this target group under IRDP. While individual artisans would continue to be covered with the supply of toolkits in traditional crafts, it is also proposed to cover the rural artisans through a cluster approach and to provide/saturate each cluster with trade-specific toolkits and related inputs in a package. In the implementation of this group cluster approach, due emphasis would be given to the formation of women artisan groups.

Availability of infrastructure facilities is an essential prerequisite for the success of IRDP activities. Substantial investments in programme infrastructure would be ensured, through a larger apportionment of funds, in consonance with the enhanced provisions of 20 per cent for IRDP infrastructure (and 25 per cent in the North Eastern States). There would be a special emphasis on infrastructure created under ISB sector with the setting up of service-cum-facility workshops at convenient places in the rural areas. These could provide common facilities in the use of machines and equipments to the rural artisans as also for repairing electrical gadgets of various types, agricultural implements, automobile parts and articles of common use. These workshops would also be set up in the tribal areas so that the process of initial value-addition to minor forest produce could be undertaken by the tribals themselves. Funding for infrastructure would also include the setting up or upgradation of technology resource centres.

Important Aspect of Infrastructural Development

Provision of marketing facilities is an important aspect of infrastructural development. In the Eighth Plan detailed guidelines were issued for setting up District Supply and Marketing Societies (DSMS) with the objective of providing integrated services to IRDP beneficiaries in the cottage and rural industries sector for the supply of raw materials, marketing of surplus products, information on technological upgradation and extension of credit support. Whereas some State Governments have taken the initiative in this regard, by and large the DSMS or similar bodies do not have much of a presence in most areas. In the Ninth Plan alternative strategies would be formulated for developing a suitable marketing infrastructure under IRDP. These would include (i) provision of transport arrangements for carrying IRDP products to rural/urban markets; (ii) introduction of insurance cover to mobile sellers; (iii) provision of better storage facilities; (iv) setting up and revamping of District Supply and Marketing Societies; (v) sale of IRDP products through Khadi & Village Industries Commission (KVIC) outlets, State Emporia etc., besides networking with DRDA showrooms/markets; (vi) setting up of quality control centres and consultancy centres which the beneficiaries could approach for advice for improving the quality and standard of their products; (vii) involvement of private sector in marketing by adoption of better packaging techniques, design, input, quality control, brand name etc.; and (viii) launching of a suitable advertisement campaign for IRDP products including organisation of exhibitions, melas. Furthermore, the potential of rural haats as a rural marketing outlet for IRDP products would be fully exploited.

There are certain areas in the country which have a very poor banking infrastructure. The areas of North-East and parts of Jammu & Kashmir fall under this category. Attention would be given to evolving innovative strategies and programmes to take care of the unbanked areas.

However, while diversifying the rural economies in high-productivity sectors, provision of adequate training facilities and upgradation of skills would be given primacy. As we have seen, the TRYSEM programme has been the weak link in the overall strategy for self-employment in rural areas. In the Ninth Plan, therefore, training would be made an integral component of IRDP. This integration would reduce to a great extent, the area-skill mismatch as the training would only be imparted in those trades/occupations which have a market potential. Yet, training would not be made compulsory as there are certain trades/activities under IRDP which do not require this input.

To strengthen the content and design of the training curriculum, the training institutes would have to constantly upgrade their syllabi in tune with the rapid changes in the job market. A basic foundation course would be a critical ingredient of the training curriculum which would make the trainees aware of simple accounting procedures, book keeping techniques and procedures in financial management, information on how to approach the banks and other financial institutions for loans, where to access the latest technology etc. Training would be given a sharper employment focus wherein the training content would be compatible with the area-specific ground realities and would be specifically imparted in those trades and activities which have a market orientation/potential and which would ensure the beneficiary sustained employment.

The quality of training would be improved. Inadequacy of proper infrastructural support has posed as a bottleneck in this effort. While the emphasis would continue to be on imparting training through the established and recognised training institutes like ITIs, Community Polytechnics, Krishi Vigyan Kendras, etc., the training infrastructure in these professionalised training institutes would be suitably strengthened. Special thrust will be given to the creation of training opportunities for women via strengthening of women ITIs, womens wings in general ITIs and women polytechnic. It is also necessary to upgrade the training skills of the trainers in the various government institutions imparting training to IRDP beneficiaries. The existing craft training centres and skill development institutes etc. would be revamped to cater to the needs of the changing situations. In those blocks where there is a concentration of unemployed youth and where there are no reputed training institutes in the vicinity, mini-ITIs should be set up, but only very selectively. Smaller private institutions and craftsmen will be engaged only where institutional support is not available.

In addition, a more effective liaison and interface between the State Governments, DRDAs and the formal/informal private sector/NGOs is required in order to identify the available employment opportunities in a region and develop training modules accordingly. Efforts would be made to establish a direct linkage between ITIs and industries in the areas where the youth could either be employed directly or could set up ancillaries to cater to the industrial demand. A similar linkage could be established between ITIs and exporters/export houses on the lines similar to gem cutting training in Gujarat. Such linkages are possible in rural hinterlands of towns/urban agglomerations. It is proposed to develop a Management Information System (MIS) through which important training institutes of relevance for IRDP throughout the country are identified and networked. This would give some idea of the areas deficient in the training infrastructure and would indicate where future investment should flow for meeting the training requirements.

With a view to ensuring that the benefits under the programme reach the more vulnerable sections of the society, the Ninth Plan would continue with the assured coverage of atleast 50 per cent for Scheduled Caste and Scheduled Tribe families, 40 per cent for women and 3 per cent for the physically handicapped. Experience in the implementation of IRDP suggests that programmes focussed on women, or in which women have played a dominant role, have performed better. The DWCRA programme has been an excellent vehicle for extending IRDP credit support for women beneficiaries in some States. The group strategy would, thereby, become the main plank for achieving the stipulated reservation for women over the next few years.

In the Ninth Plan social mobilisation will receive a special thrust. Initiative will be taken to build and strengthen the organisations of the poor with the objective of enhancing their capabilities. In this process voluntary organisations would have to play an important role in empowering the poor through advocacy, awareness generation and social mobilisation. As social animators and rural organisers they would help the poor to form SHGs in order to take advantage of the policies and programmes being implemented by the Government for their economic betterment. A voluntary organisation or a technology group could lend support to group activities by ensuring training, technological upgradation and convergence of various schemes. In addition, the DRDAs and PRIs would provide support for capacity building and provide access to credit, technology and markets to SHGs. A cadre of para professionals from within the community would be created to enhance the capability of the SHGs and help the community to access the facilities and services meant for them.

The DRDAs would be restructured in the light of the 73rd Constitutional Amendment Act, which has enhanced their area of operation, commensurate with a larger inflow of funds. The scope of the activities implemented by the DRDAs has enlarged progressively over the years from the sole implementation of rural self-employment programmes to exercising effective coordination and supervision over other Centrally Sponsored Schemes. In the revised format, the DRDAs would work under the supervision and overall control of the Zilla Parishads. The Chairman of the Zilla Parishad will be the Chairman of the DRDA Governing Body and will preside over their meetings. The organisational structure (staffing pattern) of the DRDAs would be suitably revamped and strengthened with the induction of professional cadres and technical experts. The expertise would have to be developed in the fields of credit, technology upgradation, activity-specific training and infrastructural development. To make this implementing agency effective, the staff would have to be given sufficient exposure to the complexities of poverty eradication, human resource development, gender and related issues. The DRDA's agenda for operation in the Ninth Plan, would not be limited only to the achievement of physical targets, but also to ensure the quality of the programme and realisation of intended benefits for the targeted poor beneficiaries.

WAGE EMPLOYMENT PROGRAMMES

The thrust of the Government policy is on assuring at least 100 days of employment per person per year by coordinating all the wage-employment schemes including the State schemes like the Employment Guarantee Scheme in Maharashtra. Now, with the universalisation of the EAS, both JRY and EAS are operating in all the districts/blocks of the country. At present, the JRY is being implemented through the panchayats and the EAS through the administrative apparatus. However, in consonance with the Government policy, the EAS would also be implemented through the Panchayati Raj Institutions (PRIs) in future. There is, therefore, a near complementarity of objectives and there would be a common implementing agency namely the PRIs in the case of both JRY and EAS.

Wage Employment Programmes:

- JRY will be confined to the creation of rural infrastructure according to the felt needs of the people at the village level through panchayat.
- The EAS would be the major wage-employment programme.
- In order to make EAS effective in reaching the poor and more backward areas of the country, it is proposed that funds released to States should be made as per incidence of poverty.
- Works can be taken up in blocks where there is concentration of the poor and underemployed, and workers have registered themselves for manual work.

The only major difference is that while, under JRY the allocations are made on the basis of a specified criteria, the EAS is a demand-driven scheme. In the latter case, only initial notional allocations are made and then depending on the demand for supplementary employment and actual utilisation, requests for additional allocations are made by States/Districts. Therefore, in theory there is a good case for merging these two schemes into a single wage-employment programme. Though the merger of the schemes may be operationally expedient, there would be practical constraints. Under the EAS works are taken up at the block level whenever workers register themselves for casual manual employment. Hence, it is possible that no works are taken up in many villages, particularly those which are' economically better off or where the population is sparse. In this case, these villages would be denied funds presently flowing to them under JRY. This would lead to contradictions as JRY has made the villages more independent in so far as they can take up small works according to the felt needs of the community without having to look for funds from the district administration. The JRY has also been successful in taking development to the people with planning and implementation of works according to their felt needs.

Therefore, it is proposed that the JRY would be confined to the creation of rural infrastructure at the village panchayat level. To the extent that the works would be largely labour-intensive, supplementary wage-employment will be generated and this would be monitored without the existing stipulation of a wage-material ratio of 60:40.

The EAS would be the major wage-employment programme. The funds at present flowing to the block and district levels under JRY would also be rediverted to the EAS, augmenting the resources under this scheme. However, certain precautionary measures would have to be taken in order to make the scheme effective in reaching the poor and the more backward areas of the country, where there is a concentration of the poor and underemployed. Towards this end, it is proposed that funds released to States should be made on the basis of the incidence of poverty in order to prevent the better-off States from cornering a larger share of the funds. To elaborate, in progressive States characterised by higher levels of agricultural productivity, higher per capita income and/or high literacy rates there is little real demand for casual manual work on public works. Yet, in these States demand is artificially created as, in the lean agricultural season, labour is willing to work on public works at the prevailing minimum wage rates which are relatively high. In this way, the State's specific real demand is not being addressed and in fact, there is a diversion of resources from the relatively poorer States to the more progressive ones. In such a situation, EAS would cease to be a genuine demand-driven scheme and, therefore, suitable measures would have to be taken to ensure that scarce resources reach the targeted poor and the more backward areas of the country.

Hence, as a second step, the allocations from the States to the districts will also be made on the basis of a predetermined criterion which reflects the relative backwardness of the district. Below the district level, the works can be taken up in blocks where workers have registered themselves for casual manual work. Urbanised blocks would be excluded from the purview of this scheme. No contractors would be involved. This scheme would essentially be implemented by the PRIs at the block level, but with a specific apportionment of funds to the DRDA/ZPs, for taking up inter-block and district level works. Presently, under EAS the funds are earmarked for specific activities in given proportions. Fifty per cent is earmarked for watershed-related activities. It is argued by many that it may not be fair to pre-empt the bulk of the resources for watershed-related activities being implemented in a few villages to the relative neglect of other villages in a block. In the spirit of democratic decentralised planning, the choice of activities must be left to the Panchayat Samities/Zilla Parishads which can select from the various activities/works envisaged in their block/district plans. However, at least 40 per cent of the expenditure would have to be on watershed-related activities in DPAP areas as water conservation through construction of water harvesting structures has become one of the priorities of planning, particularly in the plateau and rocky areas where water tables are falling rapidly and ground water acquifers need recharging. It may be worthwhile to spell out that the works should be labour-intensive with a 60:40 ratio of wages and materials and should lead to the creation of durable assets and rural infrastructure.

In short, EAS will be the single wage-employment programme operating throughout the country, while JRY will be confined to the creation of durable community assets at the village level. So far, under the JRY a certain quantum of funds was earmarked for Special and Innovative Projects. However, experience gained from the implementation of this component of JRY has indicated that most of the projects were not really innovative in nature but merely a replication of the existing programmes on traditional lines. Hence, there is no need to continue a separate scheme for Special and Innovative Projects.

Currently, the funds for IRDP and its allied programmes are shared between the Centre and the States in the ratio of 50:50, whereas the wage-employment programmes such as the JRY and the EAS are funded by the Centre and the States in the ratio of 80:20. It hardly needs to be emphasised that the wage-employment programmes are more popular with the State Governments on account of this cost sharing pattern. As against this, the equal cost sharing formula for self-employment schemes seems to have dampened the enthusiasm of the State Governments towards IRDP and allied programmes. States with difficult resource position find it hard to contribute their share of the outlay. In the Ninth Plan, it is proposed to alter the funding pattern under IRDP and bring it at par with the rural wage-employment programmes. In fact, the cost sharing formula for both IRDP and the wage employment programmes would be revised to a uniform ratio of 75:25 between the Centre and the States.

MILLION WELLS SCHEMES (MWS)

In view of its importance in providing a source of irrigation to the target group, the MWS would continue in the Ninth Plan to provide for maximisation of agricultural output. Simultaneously, the level of ground water would have to be maintained by the adoption of suitable recharging practices so that the small and marginal farmers can derive maximum benefits from their small holdings.

Until recently, the focus under the MWS was on the creation of employment, with the secondary objective of providing a source of irrigation.

This situation has now been modified by delinking the MWS from the JRY and repositioning it as a beneficiary-oriented scheme of irrigation for enhancing agricultural productivity levels of the small and marginal farmers.

The per capita assistance under the MWS for providing a well on the land of the small and marginal farmer ranges approximately from Rs.30,000 to Rs.35,000, though some States have reported a cost of upto Rs.1 lakh. Such a well can irrigate, on an average, one or two hectares. There have been suggestions from some States to permit the construction of field channels in conjunction with the digging of wells. This is particularly significant in view of the fact that water lifted from the wells is conveyed through open unlined earthern channels by most of the farmers, leading to conveyance losses of about 14 to 19 per cent. Necessary control and diversion structures for open line channels may be provided to avoid wastage of water. The channels could consist of structures of soil, cement or even prefabricated cement concrete.

NATIONAL SOCIAL ASSISTANCE PROGRAMME (NSAP)

The NSAP was introduced as a social security programme for the welfare of the poor households. However, it must be recognised that it is not a poverty alleviation scheme but more in the nature of a welfare programme and hence, has a limited role in the overall strategy of poverty alleviation. Moreover, the financial assistance provided to poor households in the case of old age, death of primary bread winner and maternity is in the nature of transfer payments and therefore, it should really form a part of non plan expenditure. Furthermore, the performance under this programme indicates that it has not been effective in reaching the intended beneficiaries. Moreover, several States have their own schemes of old age pension and maternity benefit for which provisions are made in the States Plans and this should continue to be within their purview, as it is a State subject. Yet, as this programme has now been introduced as a Centrally Sponsored Scheme it would continue in the Ninth Plan but would require a review thereafter.

Land Reforms

In the Ninth Plan the issue of agrarian restructuring will continue to receive the top most priority in the expectation that the States would be able to facilitate changes that would make for more efficient agriculture, leading to increases in both output and employment. This process will, in turn, contribute to the achievement of a higher rate of economic growth with social justice.

The main components of the land reform policy are the detection of ceiling surplus land and the distribution of the existing surplus land, besides tenancy reform, consolidation of holdings, providing access to the poor on common lands and wastelands, preventing the alienation of tribal lands and providing land rights to women. However, for the successful implementation of land reforms, updating of the land records, both by traditional methods and through computerisation, is an essential prerequisite. Let us elaborate.

Ceiling on Land Holdings: With the introduction of the Land Ceilings Act in 1972, the ceiling on land holdings was introduced in almost all the States with the exception of some North-Eastern States, though the ceiling limit varied depending on the quality of the land. The ceiling surplus land was to be distributed among the landless poor. In this way, land ceiling was considered an important instrument for reducing disparities in the ownership of land and as a way of increasing productivity through greater utilisation of labour. However, in practice, the extent of land declared surplus was very limited. The total area declared surplus is estimated at less than 2 per cent of the total cultivated area. Further, distribution of surplus land has been limited in several States because of institutional and legal rigidities. There are other reasons too which have led to the poor performance with regard to land ceiling. Most importantly, the ceiling surplus land continues to exist in a concealed way particularly in areas which have been covered by irrigation in the post-1972 era. Further evidence suggests that there are areas where land owners have land in excess of ceiling limits which can be mopped up if the programme for unearthing it is pursued vigorously. Also, in order to circumvent the ceiling laws, benami and furzi transfers are effected which need to be identified. There is clearly a need to detect the land falling within the ceiling limits and redistribute it. But detection and distribution of ceiling surplus land require tremendous political and administrative will.

In the wake of the economic liberalisation in certain States like Karnataka, the industry and the large farmers are being given exemption from ceiling laws without seeking the permission of the Government of India. This would certainly go against the interest of the poor as it would increase landlessness and depress agricultural wages. Hence, this issue requires close examination, before such exemptions are given.

Small farms are efficient in terms of yields and returns per unit of area. However, they are often not able to generate adequate income to sustain further investment. Therefore, there is a need for both horizontal and vertical diversification of small farms. An analysis of Indian Council of Agricultural Research (ICAR's) data of national demonstrations shows that there are large technological/yield gaps, particularly in the backward regions, which, if fully exploited, would enable the marginal and small farmers to generate incomes well above the sustenance level, subject to crop rotation, cropping pattern, yield responses of technological adoption, etc. In fact, in irrigated areas, the small and marginal farmers are viable because of their capacity to produce two or three crops a year. In the case of rainfed agriculture, a breakthrough in dryland farming is required with respect to the traditional cereal crop that can grow in these areas. Alternatively, there is a possibility of shifting to other high-value crops like fruits, vegetables, mulberry, etc. Other support systems would have to be appropriately developed to provide facilities for credit, marketing, storage and transportation of the perishable high-value crops. In fact, a change in the cropping pattern away from cereals to some high-value crops, would certainly make the small and marginal farmers viable. Also, given the increasing market demand for certain products, sericulture, horticulture and aquaculture have a great potential. In Maharashtra and West Bengal a small piece of land yields adequate returns from horticulture and acquaculture respectively. This provides further justification for ensuring that ceilings are strictly enforced as small farms can be perfectly viable in the given context.

As observed in the UNDP Human Development Report 1996, as land is redistributed from big to small farms, not only the family labour per hectare can increase sharply, so, can hired labour also. For both the reasons, the employment situation improves even for those, who remain landless after the land reform.

The main conclusion from this is that an agricultural strategy centered on small farms, rather than large, simultaneously increases the social efficiency of resource use in agriculture and improves social equity through employment creation and more equal income distribution, that small farms generate.

While there appears to be a rationale for reducing the existing ceiling limits further with a view to alleviating poverty of the growing number of rural landless poor, this is unlikely to have the support of the State Governments. Therefore, a pragmatic approach would be to strictly enforce the existing ceiling limit without permitting any attempt to circumvent it. Given the employment elasticity of agriculture in areas, which are agriculturally backward, land reforms would ensure both agricultural growth and greater employment for the rural poor. It would also provide a social status to the large number of poor. A greater transparency in the method of distribution of surplus land is possible with a greater involvement of panchayats, local communities and NGOs.

Tenancy Reforms: The policy with regard to tenancy clearly provided for conferment of ownership rights on tenants or for acquisition of ownership rights by them on payment of a reasonable compensation to the landlords. The tenancy laws of most States abolished tenancies so as to vest ownership of land with the actual tiller. Despite the legal provisions to abolish tenancy, it has continued to flourish with the existence of a large number of tenants and sub-tenants without any protection against eviction leading to insecurity among them. The continued existence of oral and concealed tenancies has led to low investments and low productivity in agriculture in several States where the implementation of tenancy reforms has been tardy. In contrast, in West Bengal, Karnataka, and Kerala, much success has been achieved in this respect. The success of 'Operation Barga' in West Bengal is well documented. This shows that conferring occupancy rights on the tenants has led to better investments in land and consequently, a higher rate of return. Hence, it is desirable that States which have a high incidence of concealed tenancy recognise this fact and record the rights of tenants. In addition, absentee landlordism has to be restrained. To the extent that large land owners leave their land fallow, either the State should take it over and lease it out on a long-term basis or the land owners should be required to lease it out on a long-term basis. Measures would have to be taken for the protection of tenants against displacement, eviction and other forms of exploitation.

A ban on tenancy was imposed in almost all the States to encourage owner cultivation and to give security of tenure to the sharecroppers and the tenants. In areas characterised by semi-feudal modes of production, where agricultural markets are not well developed, this ban is desirable with a view to protecting the tenants. It has been demonstrated that in 1972-73 in Purnea district of Bihar 40 per cent of the land was under sharecropping. This percentage has gone down and is probably around 25 per cent. The share of landholders is limited to 25 per cent of the gross produce but in practice this is perhaps illusory. Since the tenancies are oral and the sharecroppers are weak their hold on land is tenuous and they have to give to the landlord more than half of the produce. In this case, the tenants have no incentive to make long-term investments in land. The landlords also do not look upon it as a productive asset but a store of value and a reflection of their social status. Both the landlord and the sharecropper would gain if the sharecropper had the rights of cultivation in the land as he would make greater investments which would lead to higher returns. This shows that in areas where tenancy flourishes, reforms are required together with a proper implementation of laws.

On the other hand, in the "Green Revolution" areas where there is greater awareness and the markets work, freeing the lease market for land would contribute both to equity and efficiency. In these areas where the traditional arrangements exist, the tenants have been reduced to the status of an agricultural labourer with the landlord exercising considerable influence. In this case, the tenants as well as small and marginal farmers would be able to augment their operational holdings by leasing in land and, with a greater intensity of cultivation, it would lead to greater output. Clearly, agricultural tenancy should be opened up and leasing in of land permitted subject to the ceiling limits. This would activise the land market which would enhance the poor people's access to land.

Protection of Tribal Land: Despite the commitment that the tribal lands must remain with the tribals, alienation of the tribals from their land continues on a large scale due to various legal loopholes and administrative lapses. Hence, in the Ninth Plan legal provisions must be made for the prevention of alienation of tribal lands and for their restoration, not only in the notified scheduled tribe areas but also in the tribal lands in other areas. Also, the regulation of resale of the tribal lands should be made as stringent as possible.

Given that alienation is basically a consequence of economic deprivation and social discrimination it is felt, that for the tribal communities, various development programmes must be dovetailed in order to improve the income level of these people and to provide them with a basis for sustained livelihood. Even in the case of acquisition of lands of the tribals by the Government for public purposes, it should be restricted to a minimum. Encroachment of land belonging to tribals should not be permitted. Even civil courts should have no jurisdiction in the proceedings involving transfer of land of persons belonging to scheduled tribes.

Consolidation of Holdings: As already stated, consolidation of holdings has been successful in a very few States though several States have enacted legislation in this regard. In some cases, there are genuine problems in the process of consolidation including proper valuation of land, fear of eviction of small and marginal farmers who are tenants, inadequate availability of staff and lack of updated land records. Despite these constraints, consolidation of holdings makes for efficient land use and water management, leading to higher productivity and, therefore, must be enforced wherever practicable. However, in so far as there is insecurity among the tenants and small and marginal farmers, the State must ensure that their interests are protected. Consolidation operations, whenever undertaken, should be integrated with survey and settlement operations in order to avoid duplication in work and harassment to the affected person. The involvement of the Panchayati Raj Institutions should facilitate this process through a greater participation of the village people.

Government Land and Common Property Resources: In rural areas every village has common lands as well as other common property resources. These are a source of sustenance to the landless. Evidence suggests that considerable area of the government land has been taken over both by the rural poor as well as by the rural elite for agricultural and housing purposes. Unluckily, due to negligence, unsustainable overuse and excessive pressure of population on land, the productive capacity of common lands has been diminishing. However, joint forest management and watershed development programmes are schemes which can be successful in the regeneration of these common lands. Clearly, 'pattas' should be given to the rural poor in order to provide them access to a means of livelihood. There are several success stories, both in the area of joint forest management as well as watershed development by groups of people, especially women on common land, which could be replicated. These lands can also be used for providing grazing land, fodder and fuel to the poor. If the common property resources vest in the Gram Panchayat, access to it should be limited to the rural poor and the rural elite should not be allowed to encroach upon it. Similarly, the wastelands which are lying unutilised should be reclaimed and distributed among the marginal farmers and landless labourers of the area. This would provide them a source of income as also generate greater output from a variety of activities in the area of agriculture, horticulture, fodder, fuelwood and other agro-forestry. Given the paucity of 'cultivable' land, it is necessary to redistribute the wastelands and fallow land in order to provide productive employment to the poor.

In the case of land vested and in possession of the Government and/ or where a final decision is pending in a judicial court, it is suggested that such land should not be kept vacant but allotted to eligible rural poor on a short-term basis with a clear understanding that they would have to handover the land to the original landlord if the court so directs. However, legal opinion is being sought on the possibility of taking up this measure. Again in the case of Government wasteland and degraded forests, there is a growing opinion that such land should be allotted/leased to the corporate sector and for industrial and commercial purposes. As argued above, this too would adversely affect the rural poor. Access to such land should be restricted to the rural poor in order to provide them with employment.

Gender and Land Rights: So far, the strategy of land reforms has not given any cognizance to the existing gender inequalities in land inheritance laws and ceiling laws. In most regions of the country, women constitute a disproportionate number of poor. They are also more dependent on agriculture for a livelihood than men, as men shift to non-farm employment. Also, it is estimated that 20 per cent of the rural households are *de facto* female-headed. Yet, very few women have titles to land and even fewer control it. Hence, ensuring women's effective command over land will be one of the new priorities of the Ninth Plan.

Traditionally, it was accepted that agricultural land would be inherited by sons, even though in some States the inheritance law did not stipulate such a provision. Hence, it is necessary to implement the laws and also to record the rights of women, where it is legally claimed by them. Further, in so far as inheritance rights in tenancy laws are subject to specific land acts, the issues cannot be easily resolved. In some States the tenancy laws provide for devolution of tenancy to male descendants and only in their absence, women can inherit, but then also to a limited extent. Hence, it is recommended that States should amend their land and inheritance laws so as to bring agricultural land at par with other forms of property. There has been some progress in this direction in Southern and Central India.

Many of our States have improved women's access to land and landed property. States like Karnataka, Tamil Nadu and Andhra Pradesh have amended the Hindu Succession Act, 1956 to formalise issues related to women's right to property including land. Some, like Rajasthan and Madhya Pradesh have provided that issues related to property, including landed property, would be dealt with in accordance with the appropriate personal laws. However, serious anomalies continue to persist. A number of States like

U.P. Haryana, J&K, Delhi and Punjab are apparently yet to take adequate steps to provide the Constitutional/legal safeguards to women with respect to their access to land.

Additionally, it is necessary that when ceilings are fixed in relation to a "family unit", the definition of a family should be uniform across States and sons and daughters should be given equal consideration, both while assessing ceiling surplus land and in land distribution under various resettlement programmes. The pattas should be in the name of women to a larger extent. While joint pattas are better than no pattas, which do not provide the women control over it. In fact, groups of poor rural women should be given group pattas with usufructuary rights, but no right to sell individually. This group approach would enable the women to retain control over land. They could invest in the land collectively and cooperate in sharing both labour time and the returns. There are several instances of such joint management by groups of women which need to be replicated. However, to enable women to reap the benefits of land acquisition, greater access to information, credit, inputs, marketing and technologies must be provided to them.

In order to have a better data base, both the Agricultural Census and the National Sample Surveys should provide information on land ownership, land operated and land under tenancy by gender. So far, land-based statistics are recorded on a household basis. A pilot survey should be undertaken proceeding the next nationwide NSS data collection exercise.

Updating of Access to Land Records: Maintenance of up-to-date land records is crucial for effective implementation of land reforms. In recognition of this, during the Seventh Plan period, a Centrally Sponsored Scheme for Strengthening of Revenue Administration and Updating of Land Records was introduced with a view to strengthening survey and settlement organisations at the grassroots level through training, equipment, staff etc. Hence, it is imperative that States should give special emphasis to resurveys and adopt appropriate modern technologies for this, including aerial survey, photogrametic systems, global positioning system, use of scanners, digital computerised maps etc.

Further, a special drive for recording land rights of tenants and sharecroppers, as in West Bengal, must be undertaken by other States. As recommended by the Conference of Revenue Ministers held in 1997 all the existing tenancy reform legislations should be examined and a model legislation prepared.

Another Centrally Sponsored Scheme for Computerisation of Land Records was started during 1988-89 as a pilot project. But the scheme has made little progress. In fact, several factors including power shortage, lack of trained staff and bureaucratic procedures involved at various stages, have constrained the progress of computerisation of land records. Despite these problems, computerisation must be given a high priority and solutions found for the operational problems. The updating of land records can be expedited even without computerisation through the involvement of panchayat institutions and the local revenue functionaries. The village-level revenue functionaries should be placed under the control of the Gram Panchayats, though the appellate jurisdiction should continue with the tehsildar. At the district level, the land revenue system must work under the Zilla Parishad. The 30 per cent representation for women in PRIs should help the cause of women in so far as recording of rights of women in land is concerned.

Moreover, steps have to be taken to bring about greater transparency in the administration of land records, with access to information regarding land holdings on demand by any individual. Copies of land records, including record of rights, field books/field map, Land Pass Books as also mutation statements, status of land and jamabandi register should be accessible and copies provided on payment of a fee.

ANTI-POOR LAWS AND POLICIES

Over time, the expenditure on the various poverty alleviation programmes has increased significantly. It is true that these programmes play a vital role in ensuring that the poor are able to sustain themselves, particularly during the lean seasons, through generation of additional employment and incomes. However, it is becoming increasingly clear that spending money is not the only way of ameliorating the conditions of the poor. Scant attention has been paid to the role of non-monetary policies and institutional arrangements which affect the lives of the common people, especially the poor. There are several laws and policies which are anti-poor. The knowledge about them is scattered and often anecdotal. However, some examples may illustrate the nature of the problem.

For tribals and forest dwellers minor forest produce obtained from forest and common property resources is a major source of livelihood. However, the poor tribals are facing problems in gathering this minor forest produce due to diversion of non-timber forest produce to the industry. Bamboo forests are being readily leased to the paper industries regardless of the provision in the Forest Policy 1988 which stipulates that the forest dwellers have the first charge on the forest produce. Rural artisans, who make products out of bamboo, face shortage of raw material as the forest departments allot them green bamboo in rationed quantities. This practice has been observed in Karnataka and Orissa. In some cases, forest land is leased out to private industry for long periods. This too adversely affects the poor.

The forest department also issues licences to women for gum collection from babool trees and compels them to sell their entire collection to the forest department at a prescribed price which is one third to one fourth of the market price. In some States, tribals can collect hill brooms but cannot process these into a broom nor can they sell the collected items in the open market. Thus, the poor are prevented from making value addition through processing and are denied the right to get the best price for their produce. Licensing activities for the poor in rural areas are cumbersome and time consuming. To set up a charcol kiln one requires four to five types of permission.

The cattle flayers in UP have no legal control over their own produce *i.e.* the hides they flay from the naturally fallen animals. The Zilla Parishads award contracts either to individuals or to cooperatives for collection and storage of hides. Hence, the flayers are not able to sell in the open market but are forced to accept the terms made by the contractors. Again in Bihar, auction ferry right is made in favour of Zamindars and contractors who collect tolls and taxes from the poor who have to cross the river to sell their produce. In addition, those who are fishing or plying boats across the river have to pay taxes to the Zamindars. Normally, access to, and use of, the river should be free to farmers, labourers, petty traders etc. In Assam, the fishery laws are such that the fishermen do not benefit from them. A sector of the river is leased out to "Mohildhars" who auction the fish and only 1/3rd is given to fishermen who actually catch the fish. These are ways in which common property resources are being used in a way that are prejudicial to the poor.

The rural poor do not have a level playing field as the market for their finished products are dominated by a single trader or cartel of traders operating in the area. In the market, the terms of trade are against them. Neither the public nor the private sector has encouraged the growth of poor people's organisations but, instead, has either tried to suppress or control them. Cooperatives could have become the real supporters of the poor and needy but the process for their formation and registration is very cumbersome. Where cooperatives of poor people have been successful such as womens' beedi cooperatives, they have been taken over by private companies. Similarly, the Government has often interfered with the independence of the cooperatives by putting government operators incharge of them.

Such examples abound. An exercise has already been done to document some of these laws and policies. In the Ninth Plan, an attempt will be made to initiate the process of identifying anti-poor laws/policies, Statewise. These would be brought to the notice of the policy makers, local governments and NGOs so that these may be suitably modified and/or repealed in the interest of the poor.

INTEGRATION OF POVERTY ALLEVIATION PROGRAMMES WITH SECTORAL PROGRAMMES

It is necessary to recast the special employment programmes with a view to making them more effective in meeting not only the short-term objective of providing temporary work, but also in building up the productive capacity of individuals/areas which, in turn, would make for greater employment on a more sustainable basis. The focus would have to be on agriculture and allied activities, besides rural non-farm sectors and services, which have a high employment elasticity. This would require a high degree of convergence among the various poverty alleviation programmes (PAPs), area development programmes and sectoral schemes, within a district plan based on the physical and human endowments of the area, the felt needs of the people and the total financial resources available. Using scientific methods, remote-sensing agencies at the Centre and State level would be asked to provide detailed maps showing land, water and other physical resources of the area, with the aid of photogrammetry and satellite imagery. The detailed maps would then be scrutinised to identify all possible watersheds. Planning along watershed lines would ensure minimum surface run-off, thus conserving water from rainfall. Viable activities in agriculture and allied sectors would have to be selected. Agro-processing activities linked with the cropping pattern, village and small industries with growth potential and other infrastructural gaps would also have to be identified and prioritised. Within this framework, poverty alleviation programmes would be integrated with area development programmes within a developmental plan at the district level.

So far, there has been a complete dichotomy between various sectoral as well as poverty alleviation programmes that have been planned and implemented by the concerned line departments. The Government has recognised this dichotomy but greater efforts have to be made to effect the convergence in practice. As a start, an attempt has been made to integrate DPAP and DDP, EAS and the Integrated Wasteland Development Programme (IWDP), all being implemented by a single Ministry. Watersheds are to be constructed and associated works of drainage, land development and terracing undertaken. Also, afforestation, agro-based and horticultural development, pasture development, crop demonstration for popularising new crops/ varieties and upgrading of common property resources are being taken up. Clearly, there is scope for integrating other sectoral programmes of soil and water conservation, forestry, minor irrigation, animal husbandry, agriculture and other departments, funds from which flow from sectoral heads to the district level. Once the area plans are prepared, dovetailing of funds would not pose a problem.

INSTITUTIONAL MECHANISM FOR DELIVERY

Consequent to the 73rd Constitutional Amendment Act, State Governments have enacted enabling legislations providing for local bodies at the village, intermediate and district levels. In almost all the States with the exception of Bihar the PRIs have been duly constituted. The State Governments are required to endow these PRIs with the requisite financial and administrative powers to enable them to function as institutions of self-government. The PRIs would be responsible for preparing the plans for economic development and social justice through the District Planning Committee and for implementing them, in respect of the items listed in the Eleventh Schedule of the Constitutional Amendment. This process of devolution is at various levels of operationalisation for individual States. It is expected, that in the Ninth Plan, the States would devolve funds on the panchayats both from the Consolidated Funds of the States and the allocations made by the Central Government for Centrally Sponsored Schemes. In addition, the panchayats have to be given their own revenue raising powers, as per the recommendations of the State Finance Commissions. It is expected that village level plans would be prepared, based on the felt needs of the people as articulated in the Gram Sabha meetings. These plans would be incorporated into the intermediate level plans and finally merged into a district plan. This district plan would then enable the dovetailing of funds from the various sectoral poverty alleviation and area development programmes. In this way, development planning would begin from below reflecting the aspirations of the people within the constraints of the available physical and financial resources. Voluntary organisations would also play an enhanced role especially as facilitators and social animators in bringing about greater awareness through advocacy. They would also help the poor to form self-help groups with the objective of improving their economic status through concerted action. In this way the PRIs, the voluntary organisations and the community would work in tandem to bring about greater development at the local level and consequent reduction in poverty levels.

URBAN POVERTY ALLEVIATION

Poverty reduction is an important goal of the urban policy. Urban growth is a result of (1) natural increase in population (2) net migration from rural areas to urban areas and (3) reclassification of towns. The common notion that migration largely fuels urban growth is only partially correct. Therefore, it is necessary to view urban poverty as distinct from rural poverty and not as mere transfer of rural poverty into urban areas.

Urban poverty leads to : (a) proliferation of slums and bustees; (b) fast growth of the informal sector; (c) increasing casualisation of labour; (d) increasing pressure on civic services; (e) increasing educational deprivation and health contingencies.

The Urban Poverty Alleviation Programmes (UPAPs) which were in operation during eighth plan are as follows:

NEHRU ROZGAR YOJANA (NRY)

In order to alleviate the conditions of urban poor, a Centrally Sponsored programme-Nehru Rozgar Yojana-was launched at the end of the Seventh

Five Year Plan (October 1989) with the objective of providing of employment to the urban unemployed and underemployed poor. The Central Government indicated its overall contribution while the essential task of identifying, earmarking and coordinating the relevant sectoral inputs was undertaken by the State Governments. The NRY consisted of three schemes namely (i) the Scheme of Urban Micro Enterprises (SUME); (ii) the Scheme of Urban Wage Employment (SUWE); and (iii) the Scheme of Housing and Shelter Upgradation (SHASU). During the Eighth Plan, 92% of the available funds were utilised and but for the shortfall in the number of dwelling units upgraded/in progress under SHASU, the targets have been achieved under all the other schemes.

URBAN BASIC SERVICES FOR THE POOR (UBSP)

The UBSP Programme was implemented as a Centrally Sponsored Scheme during the Eighth Five Year Plan with the specific objectives of effective achievement of the social sector goals; community organisation, mobilisation and empowerment; and converence through sustainable support system. The expenditure on the Programme was being shared on a 60:40 basis between the Central and the State Governments and UTs (with legislatures). Further, the per capita expenditure on any slum pocket is Rs.75/-in the first year and Rs.50/-from the second year onwards after the basic infrastructure is developed. The UBSP was targetted to cover 70 lakh urban poor beneficiaries in 500 towns during the Eighth Plan period. The Programme has achieved the physical target of 70 lakh beneficiaries during the Eighth Plan period in 350 towns. Against the release of the Central share of Rs.8090 lakh, the release of the State share was Rs.3439.64 lakh. As on 31.03.1997, 353 towns and 4993 slum pockets have been selected for coverage and 75 lakh beneficiaries have been covered.

PRIME MINISTER'S INTEGRATED URBAN POVERTY ERADICATION PROGRAMME (PM IUPEP)

Recognising the seriousness and complexity of urban poverty problems, especially in the small towns where the situtation is more grave due to lack of resources for planning their environment and development, the PMI UPEP was launched in November, 1995. The PM IUPEP was a Rs.800 crore scheme approved for the period up to the year 2000. Programme was applicable to all Class II urban agglomerations with a population ranging between 50,000 and 1 lakh subject to the condition that elections to local bodies have been held. The Programme was being implemented on a wholetown/ project basis extending the coverage to all the targetted groups for recuring a visible impact. During 1995-96 and 1996-97, Rs.176.40 crore were released to the States/UTs. Most of the States are in the preparatory stages of the Programme, such as house-to-house survey, patial mapping, need assessment, developing lternative project reports, building community structures etc., which take quite some time. The physical achievements as reported by the States are as under:

- House-to-house survey has been completed in 213 towns.
- Town-wise project reports have been prepared for 229 towns.
- Under the self-employment component, 20775 applications have been forwarded to banks, out of which 3080 cases have been approved.
- Under the Shelter Upgradation Component, 10386 applications have been forwarded to banks/HUDCO, out of which 4743 cases have been approved by HUDCO.
- As many as 8382 Neighbourhood Groups, 1200 Neighbourhood Development Committees and 444 Thrift and Credit Societies have been formed.

The performance of the UPAPs in the Eighth Five Year Plan reveals:

- that although urban poverty is no less severe than rural poverty, the priority accorded to alleviation of urban poverty is low as the common perception is that urban poverty is a transfer of rural poverty into urban areas;
- that UPAPs are highly fragmented and have overlapping objectives and strategies.
- that integration of UPAPs with sectoral development and area development programmes has been overlooked.
- that the role of voluntary organisations and community based organisation in planning and implementation of UPAPs is on the periphery.

The Planning Commission's (Modified Expert Group) estimates of urban poverty show a decline in the percentage of urban poor from 38 percent (1987-88) to 32.36 percent (1993-94). As per the Modified Expert Group the number of urban poor is 763.37 lakh in 1993-94. The Hashim Committee, which was set up to review and rationalise Centrally Sponsored Schemes for poverty alleviation and employment generation, had gone into the question of rationalisation of the existing poverty alleviation programmes and recommended that:-

The self-employment component of NRY and PMIUPEP be combined into a single programme valid for all the urban areas all over the country;

- The Planning Commission's (modified Expert Group) estimates of urban poverty show a decline in the percentage of urban poor from 38 percent (1987-88) to 32.36 percent (1993-94). As per the Modified Expert Group the number of urban poor is 763.37 lakh in 1993-94.
- The Hashim Committee, which was set up to review and rationalise Centrally Sponsored Schemes for poverty alleviation and employment generation, had gone into the question of rationalisation of the existing poverty alleviation programmes.

- Based on the recommendation of the Committee the Swarna Jayanti Shahari Rozgar Yojana (SJSRY) has been launched with effect from 1.12.97 and NRY, PMIUPEP and UBSP have been phased out.
- The programme has two sub-schemes namely, (a) Urban Self – Employment Programme and (b) Urban Wage Employment Programme. The self-employment and wage employment components of the NRY and PMIUPEP have been re-organised under this single programme. The shelter upgradation components of both NRY and PMIUPEP has been merged with the National Slum Development Programme.

The urban wage employment component as well as the physical infrastructure development component under the NRY and the PMIUPEP be merged and be made applicable to all the urban areas with a population less than 5 lakhs. This component may be separated from the self-employment component as a separate scheme with a distinct identity; and

The shelter upgradation/housing component under NRY and PMIUPEP be retained either as a separate scheme or merged with the Slum Development/ Basic Services Schemes operating at present.

THE SWARNA JAYANTI SHAHARI ROZGAR YOJANA (SJSRY)

In pursuance of the above recommendations, during the Ninth Plan it is proposed to launch the Swarna Jayanti Shahari Rozgar Yojana (SJSRY) and phase out NRY, PMIUPEP and UBSP. The SJSRY is to be a Centrally Sponsored Scheme applicable to all the urban areas with expenditure to be shared in ratio 75:25 between the Centre and States/UTs.

The programme has become operational on December 1, 1997. This programme would have two sub-schemes, namely, (i) UrbanSelf-Employment Programme and (ii) Urban Wage Employment Programme. The self-employment and wage employment components of the existing NRY and PMIUPEP have been reorganised under this single programme. Further,the shelter upgradation components of both NRY and PMIUPEP will be merged with the National Slum Development Programme.

The Swarna Jayanti Shahari Rozgar Yojana (SJSRY) seek to provide gainful employment to the urban unemployed or underemployed poor by encouraging the setting up of self-employment ventures or provision of wage employment. This programme will rely on the creation of suitable community structures on the UBSP pattern and delivery of inputs under this programme will be through the medium of urban local bodies and similar community institutional structures.

The Swarna Jayanti Shahari Rozgar Yojana rests on the foundation of community empowerment. Towards this end, community organisations like Neighbourhood Groups (NHGs), Neighbourhood Committees (NHCs) and Community Development Societies (CDSs) will be set up in the target areas based on the UBSP pattern. The CDSs will be the focal point for purposes of identification of beneficiaries, preparation of applications, monitoring of recovery and generally providing whatever other support is necessary to the programme. The CDSs will also identify viable projects suitable for that particular area.

These CDSs may also set themselves up as Thrift and Credit Societies to encourage community savings, as also other group activities. A maximum expenditure at the rate of Rs.100 per member for the first year and Rs.75 per member for each subsequent year will be allowed for activities connected with the CDSs.

THE URBAN SELF EMPLOYMENT PROGRAMME (USEP)

This programme will have three distinct components:-

i. assistance to individual urban poor beneficiaries for setting up gainful self-employment ventures.
ii. assistance to groups of urban poor women for setting up gainful self-employment ventures. This sub-scheme may be called "The Scheme for Development of Women and Children in the Urban Areas (DWCUA)".
iii. training of beneficiaries, potential beneficiaries and other persons associated with the urban employment programme for upgradation and acquisition of vocational and entrepreneurial skills.

Coverage

i. The programme will be applicable to all urban towns in India.
ii. The programme will be implemented on a whole town basis with special emphasis on urban poor clusters.

Target Groups

i. The programme will target the urban poor,i.e those living below the urban poverty line, as defined from time to time.
ii. Special attention will be given to women, persons belonging to Scheduled Castes/Tribes, disabled persons and other such categories as may be indicated by the Government from time to time. The percentage of women beneficiaries under this programme will not be less than 30 per cent. The SCs and STs must be benefited at least to the extent of their proportion in the local population. A special provision of 3 percent will be reserved for the disabled under this programme.
iii. There will not be any minimum educational qualification set for beneficiaries under this programme. However, to avoid an overlap with the PMRY scheme, for self-employment component, this scheme will not apply to beneficiaries educated beyond the IX

standard. As regards the wage employment component, there will be no restrictions of educational qualifications whatsoever.

iv. A house-to-house survey for identification of genuine beneficiaries will be undertaken. Non-economic parameters will also be applied to identify the urban poor in addition to the economic criteria of the urban poverty line.

All other conditions being equal, women beneficiaries belonging to women-headed households will be ranked higher in priority than other beneficiaries.

SELF-EMPLOYMENT THROUGH SETTING UP MICRO-ENTERPRISES AND SKILL DEVELOPMENT

To avoid duplication with the ongoing Prime Minister's Rozgar Yojana (PMRY), this component of SJSRY is confined to the below poverty line beneficiaries who have no education upto ninth standard, with emphasis on those accorded a higher priority on the basis of the non-economic critera. The maximum unit cost will be Rs.50,000 and the maximum allowable subsidy will be 15 percent of the project cost, subject to a limit of Rs.7,500. The beneficiary is required to contribute 5 percent of the project cost as margin money. In case a number of beneficiaries decide to jointly set up a project, such project will be eligible for a subsidy which will be equal to the total of the permitted subsidy per person as per the above criteria. In this case too the provision relating to 5 percent margin money per beneficiary will apply. The overall project cost, which can be permitted, will be the total of the individual project cost allowable per beneficiary.

Skill development through appropriate training is another element of this programme. The unit cost allowed for training will be Rs.2000 per trainee, including material cost, trainers' fees, other miscellaneous expenses to be incurred by the training institution and the monthly stipend to be paid to the trainee. The total training period for skill upgradation may vary from two to six months, subject to a minimum of 300 hours. Infrastructural support may also be provided to the beneficiaries setting up micro-enterprises in relation to marketing of their products etc.

DEVELOPMENT OF WOMEN AND CHILDREN IN URBAN AREAS (DWCUA)

This scheme is distinguished by the special incentive extended to urban poor women who decide to set up self-employment ventures as a group as opposed to individual effort. Groups of urban poor women will take up an economic activity suited to their skill, training, aptitude and local conditions. Besides generation of income, this group strategy will strive to empower the urban poor women by making them independent as also providing a facilitating atmosphere for self-employment. To be eligible for subsidy under this scheme, the DWCUA group should consist of at least 10 urban poor women.

FINANCIAL PATTERN

The DWCUA group society will be entitled to a subsidy of Rs.1,25,000 or 50% of the cost of project whichever is less. Where the DWCUA group sets itself up as a Thrift and Credit Society, in addition to its other entrepreneurial activity, the group/Thrift and Credit society will also be entitled to a lump sum grant of Rs.25,000 as a revolving fund at the rate of Rs.1000 maximum per member. This revolving fund will be available to a simple Thrift and Credit Society also even if the society is not engaged in any project activity under DWCUA.

THE URBAN WAGE EMPLOYMENT PROGRAMME (UWEP)

This programme seeks to provide wage employment to beneficiaries living below the poverty line within the jurisdiction of urban local bodies by utilising their labour for construction of socially and economically useful public assets. This programme will apply to urban local bodies, the population of which is less than 5 lakhs as per the 1991 Census.

The material-labour ratio for works under this programme will be maintained at 60:40. The prevailing minimum wage rate, as notified from time to time for each area, will be paid to the beneficiaries under this programme.

This programme will be dovetailed with the State sector EIUS scheme as well as the National Slum Development Programme (NSDP). This programme is not designed to either replace or substitute the Environmental Improvement Of Urban Slums (EIUS), the NSDP, or any other State sector schemes.

Project Administration

At the community level, a Community Organiser(CO) will be appointed for about 2000 identified families. The Community Organiser should, as far as practicable, be a woman.

At the town level, there will be an Urban Poverty Eradication Cell under the charge of a Project Officer. The Project Officer will be responsible for coordinating the activities of all the CDSs and COs.

At the district level, the State Government will constitute a District Urban Development Agency (DUDA) with an officer designated as the District Project Officer.

At the State level, there will be a State Urban Development Authority (SUDA), which will be headed by a full-time senior officer of the State Government. The SUDA will be designated as the State Nodal Agency for urban anti poverty programmes.

At the national level, the Department of Urban Employment and Poverty Alleviation will be the nodal department. The programme will be monitored and overseen by the UPA Division. The Ninth Plan, in particular, needs to focus on the effective implementation and management of urban poverty alleviation programmes (UPAPs). Targetting of urban poor is essential for the success of any anti-poverty programme. Therefore, the UPAPs should be designed to accomodate flexibility in operation. The key components of the Ninth Plan strategy would be:

- effective targetting using either a self-identification system or a selection process by institutions responsible for financing and managing UPAPs as faulty targetting will enhance income inequalities.
- a "bottom up" approach should be adopted for UPAPs which should be routed through community based organi-sations. There should be effective participation of community institutions for extending loans and for aiding loan recoveries made under UPAPs.
- since the problem of urban poverty is a manifestation of the higher incidence of marginal and low income employment in the informal sector, it is essential to upgrade informal sector occupations. Hence there is a need for emphasis on generation of self-employment in processing and services sector, improving the acccess to technology and credit and above all improving the general legal and physical environment which governs the working of the informal sector.

The policy framework adopted by the Government of India also mentions that "a frontal attack on poverty is an important element in development policy. This is the main rationale for anti-poverty programmes".

5

Urban Agriculture and Farming

INTRODUCTION

Urban agriculture is the practice of cultivating, processing and distributing food in, or around (peri-urban), a village, town or city. Urban farming is generally practiced for income-earning or food-producing activities though in some communities the main impetus is recreation and relaxation. Urban agriculture contributes to food security and food safety in two ways: first, it increases the amount of food available to people living in cities, and, second, it allows fresh vegetables and fruits and meat products to be made available to urban consumers.

A common and efficient form of urban agriculture is the biointensive method. Because urban agriculture promotes energy-saving local food production, urban and peri-urban agriculture are generally seen as sustainable practices. The recognition of environmental degradation within cities through the relocation of resources to serve urban populations has inspired the implementation of different schemes of urban agriculture across the developed and developing world. From historic models such as Machu Picchu to designs for new productive urban farms, the idea of locating agriculture in the city takes on many characteristics.

HISTORY OF URBAN AGRICULTURE

Community wastes were used in ancient Persia to feed urban farming. In Machu Picchu water was conserved and reused as part of the stepped architecture of the city and vegetable beds were designed to gather sun in order to prolong the growing season. Victory gardens sprouted during WWI, WWII and were fruit, vegetable, and herb gardens in US, Canada, and UK. This effort was undertaken by citizens to reduce pressure on food production that was to support the war effort. Community gardening in most communities are open to the public and provide space for citizens to cultivate plants for food or recreation. A community gardening programme that is well-established is Seattle's P-Patch.

URBAN AGRICULTURAL FACTS

- 50% of the world's population lives in cities.
- 800 million people are involved in urban agriculture world-wide and contribute to feeding urban residents.
- Low income urban dwellers spend between 40% and 60% of their income on food each year.
- By 2015 about 26 cities in the world are expected to have a population of 10 million or more. To feed a city of this size – at least 6000 tonnes of food must be imported each day.

PERSPECTIVES ON URBAN AGRICULTURE

Resource and Economic

The Food and Agriculture Organization of the United Nations (FAO), has defined urban agriculture as:

" [A]n industry that produces, processes and markets food and fuel, largely in response to the daily demand of consumers within a town, city, or metropolis, on land and water dispersed throughout the urban and peri-urban area, applying intensive production methods, using and reusing natural resources and urban wastes to yield a diversity of crops and livestock."

The definition of urban agriculture as an industry that responds to the nutritional demands of a city, from within that city, with the use and reuse of that city's resources while acknowledging economic and resource use does not reconcile aspects of regional health, food security, and application of grassroots organizations. (This definition is based on the work of Luc Mougeot of the International Development Research Centre and used in technical and training publications by UN-HABITAT's Urban Management Programme, FAO's Special Programme for Food Security, and international agricultural research centres, such as CIRAD.)

Environmental

The Council on Agriculture, Science and Technology, (CAST) is an international consortium of scientific and professional societies based in Ames Iowa that compiles and communicates credible science-based information to policy makers, media, private sector, and the public. CAST defines urban agriculture to include aspects of environmental health, remediation, and recreation, "Urban agriculture is a complex system encompassing a spectrum of interests, from a traditional core of activities associated with the production, processing, marketing, distribution, and consumption, to a multiplicity of other benefits and services that are less widely acknowledged and documented. These include recreation and leisure; economic vitality and business entrepreneurship, individual health and well-being; community health and well being; landscape beautification; and environmental restoration and remediation." Modern planning and design initiatives are more responsive to this model of urban agriculture because it fits within the current scope of sustainable design. The definition allows for a multitude of interpretations across cultures and time. Frequently it is tied to policy decisions to build sustainable cities.

Food Security

Access to nutritious food is another perspective in the effort to locate food and livestock production in cities. With the tremendous influx of world population to urban areas, the need for fresh and safe food is increased. Community Food Security Coalition (CFSC) defines food security as, "all persons in a community having access to culturally acceptable, nutritionally adequate food through local, non-emergency sources at all times."

THE EMERGING IMPORTANCE OF URBAN AGRICULTURE

Economic

- UPA (urban and peri-urban agriculture) expands the economic base of the city through production, processing, packaging, and marketing of consumable products. This results in an increase in entrepreneurial activities and the creation of job opportunities, as well as in food costs reduction and products of better quality.
- UPA represents an important opportunity for women to be part of the informal economy of a city. Farming and selling activities can be combined more easily with household tasks and child care.
- UPA provides employment, income, and access to food for urban populations, which together contributes to relieve chronic and emergency food insecurity. Chronic food insecurity refers to less affordable food and growing urban poverty, while emergency food insecurity relates to breakdowns in the chain of food distribution. UPA plays an important role in making food more affordable and in providing emergency supplies of food. Research into market values for produce grown in urban gardens has attributed to a community garden plot a median yield value of between approximately \$200 and \$500 (US, adjusted for inflation). In a community gardening programme as well-established as Seattle's P-Patches, this can account for up to 1.25 million dollars of produce cultivated annually.

Social

Social benefits that have emerged from urban agricultural practices are; better health and nutrition, increased income, employment, food security within the household, and community social life. UPA can be seen as a means of improving the livelihood of people living in and around cities. Taking part in such practices is seen mostly as informal activity, but in many cities where inadequate, unreliable, and irregular access to food is an occurring problem, urban agriculture has been a positive response to tackling food concerns. Households and small communities take advantage of vacant land and contribute not only to their household food needs but also the needs of their resident city.

The CFSC states that, "Community and residential gardening, as well as small-scale farming, save household food dollars. They promote nutrition and free cash for non-garden foods and other items"

This allows families to generate larger incomes selling to local grocers or to local outdoor markets, while supplying their household with proper nutrition of fresh and nutritional produce.

Some community urban farms can be quite efficient and help women find work, who in some cases are marginalized from finding employment in the formal economy. Studies have shown that participation from women have a higher production rate, therefore producing the adequate amount for household consumption while supplying more for market sale.

Due to the fact that most UA activities are conducted on vacant municipal land, there have been rising concerns about the allocation of land and property rights. The IDRC and the FAO have published the Guidelines for Municipal Policymaking on Urban Agriculture, and are working with municipal governments to create successful policy measures that can be incorporated in urban planning.

Including UA in local plans and as proper land use will continue to help impoverished communities gain a better well-being while fighting urban poverty. Localized food production in urban and peri-urban areas contributes to local economies by creating jobs and producing valuable products.

Some researchers indicate that unemployed populations in large cities and suburban towns would decrease if put to work by local food movements. Schools have foreseen the asset of local food production and are beginning to incorporate agricultural sections in their curricula and present it as a career opportunity. Urban agricultural projects are beginning to open a new labor market in areas that have been negatively affected by industrial outsourcing of jobs.

Energy Efficiency

The current industrial agriculture system is accountable for high energy costs for the transportation of foodstuffs. The average conventional produce item travels 1,500 miles, using, if shipped by tractor-trailer, one gallon of fossil fuel per hundred pounds. The energy used to transport food is decreased when urban agriculture can provide cities with locally-grown food.

Quality of Food

Although the taste of locally grown food is subjective, many participants in the urban agriculture movement report that they prefer the taste of local agricultural products, or organic food, to that of industrial food production. Also, urban agriculture supports a more sustainable production of the food that tries to decrease the use of harmful pesticides that result in agricultural runoff. Urban and local farmers also eliminate the need for preservatives, as their products do not need to travel long distances.

Economy of Scale

Using high-density urban farming, as for instance with vertical farms or stacked greenhouses, many environmental benefits can be achieved on a city-wide scale that would be impossible otherwise. These systems do not only provide food, but also produce potable water from waste water, and can recycle organic waste back to energy and nutrients. At the same time, they can reduce food-related transportation to a minimum while providing fresh food for large communities in almost any climate.

IMPLEMENTATION OF URBAN AGRICULTURE

COMMUNITY-BASED INFRASTRUCTURE

Creating a community-based infrastructure for urban agriculture means establishing local systems to grow and process food and transfer it from farmer (producer) to consumer.

To facilitate food production, cities have established community-based farming projects. Some projects have collectively-tended community farms on common land, much like that of eighteenth-century Boston Common. One such community farm is the Collingwood Children's Farm in Melbourne, Australia. Other community garden projects use the allotment garden model, in which gardeners care for individual plots in a larger gardening area, often sharing a tool shed and other amenities. Seattle's P-Patch gardens use this model, as did the South Central Farm in Los Angeles. Independent urban gardeners also grow food in individual yards and on roofs. Roof gardens allow for urban dwellers to maintain green spaces in the city without having to set aside a tract of undeveloped land. There are a growing number of projects worldwide that seek to enable cities to become 'continuous productive landscapes' through the networked cultivation of vacant urban land and temporary or permanent 'kitchen gardens'. Food processing on a community level has been accommodated by centralizing resources in community tool sheds and processing facilities for farmers to share. The Garden Resource Program Collaborative based in Detroit has cluster tool banks. Different areas of the city have toolbanks where resources like tools, compost, mulch, tomato stakes, seeds, and education can be shared and distributed with the gardeners in that cluster. Detroit's Garden Resource Program Collaborative also strengthens their gardening community by providing to their members transplants; education on gardening, policy, and food issues; and by building connectivity between gardeners through workgroups, potlucks, tours, field trips, and cluster workdays.

Farmers' markets, such as the farmers' market in Los Angeles, provide a common land where farmers can sell their product to consumers. Large cities tend to open their farmers markets on the weekends and one day in the middle of the week. For example, the farmers' market of Rue Richard Lenoir in Paris, France, is open on Sundays and Thursdays. However, to create a consumer dependency on urban agriculture and to introduce local food production as a sustainable career for farmers, markets would have to be open regularly. For example, the Los Angeles Farmers' Market is open seven days a week and has linked several local grocers together to provide different food products. The market's central location in downtown Los Angeles provides the perfect interaction for a diverse group of sellers to access their consumers.

Havana, Cuba

Due to the shortage of fuel and therefore severe deficiencies in the transportation sector a growing percentage of the agricultural production takes place in the so-called urban agriculture. In 2002, 35,000 acres (140 km^2) of urban gardens produced 3.4 million tons of food. In Havana, 90% of the city's fresh produce come from local urban farms and gardens. In 2003, more than 200,000 Cubans worked in the expanding urban agriculture sector.

Mumbai, India

Economic development in Mumbai brought a growth in population caused mainly by the migration of laborers from other regions of the country. The number of residents in the city increased more than twelve times in the last century. Greater Mumbai, formed by City Island and Salsette Island, is the largest city in India with a population of 16.4 million, according to data collected by the census of 2001. Mumbai is one of the densest cities in the world, 48,215 persons per km^2 and 16,082 per km^2 in suburban areas. In a scenario like this, urban agriculture seems unlikely to put in practice since it must compete with real estate developers for the access and use of vacant lots. Alternative farming methods have emerged as a response to scarcity of land, water, and economic resources employed in UPA.

Dr. Doshi's innovative techniques on farming: Dr. Doshi's city garden methods are revolutionary for being appropriate to apply in reduced spaces as terraces and balconies, even on civil construction walls, and for not requiring big investments in capital or long hours of work. His farming practice is purely organic and is mainly directed to domestic consumption.

His gardening tools are composed of materials available in the local environment: sugarcane waste, polyethylene bags, tires, containers and cylinders, and soil. The containers and bags (open at both ends) are filled with the sugarcane stalks, compost, and garden soil, which make possible the use of minimal quantity of water if compared to open fields. Dr. Doshi states that solar energy can replace soil in cities. He also recommends the idea of chainplanting, or growing plants in intervals and in small quantities rather than at once and in large amounts. He has grown different types of fruit such as mangos, figs, guavas, bananas, and sugarcane stalks in his terrace of 1,200 sq ft (110 m^2) in Bandra. The concept of city farming developed by Dr. Doshi consumes the entire household's organic waste. He subsequently makes the household self-sufficient in the provision of food: 5 kg of fruits and vegetables are produced daily for 300 days a year.

City farm at Rosary High School, Dockyard Road Mumbai: The main objectives of this pilot project were to promote economic support for street children, beautify the city landscape, supply locally produced organic food to urban dwellers (mainly those residing in slums), and to manage organic waste in a sustainable city. The project was conducted in the Rosary School, in Mumbai, with the participation of street children during 2004. A city farm was created in a terrace area of 400 sq ft (37 m^2). The participants were trained in urban farming techniques. The farm produced vegetables, fruits, and flowers. The idea was spread the concept of city farm to other schools in the city.

Mumbai Port Trust (MBPT): The central kitchen distributes food to approximately 3,000 employees daily, generating important amounts of organic disposal. A terrace garden created by the staff recycles ninety percent of this waste in the production of vegetables and fruits. Preeti Patil, who is the catering officer at the MBPT explains the purpose of the enterprise:

"Mumbai Port Trust has developed an organic farm on the terrace of its central kitchen, which is an area of approximately 3,000 sq ft (280 m^2). The activity of city farming was started initially to dispose of kitchen organic waste in an ecofriendly way. Staff members, after their daily work in the kitchen, tend the garden, which has about 450 plants."

Bangkok, Thailand

In early 2000, urban gardens were started under the direction of the NGO, Thailand Environment Institute (TEI), to help achieve the Bangkok Metropolitan Administrations (BMA) priority to 'green' Thailand. With a population of 12 million and 39% of the land in the city vacant due to rapid expansion of the 1960's-80's Bangkok is a test bed for urban gardens centered on community involvement. The two urban gardens initiated by TEI are in Bangkok Noi and Bangkapi and the main tasks were stated as:

- Teach members of the communities the benefits of urban green space.

- Create the social framework to plan, implement, and maintain the urban green space.
- Create a process of method to balance the needs of the community with the needs of the larger environmental concerns.

While the goals of the NGO are important in a global context, the community goals are being met through the work of forming the urban gardens themselves. In this sense the creation, implementation, and maintenance of urban gardens is highly determined by the desires of the communities involved. However, the criteria by which TEI measured their success illustrates the scope of benefits to a community which practices urban agriculture. TEI's success indicators were:

- Establishing an Urban Green Plan
- Community Capacity Building
- Poverty Reduction
- Links with Government
- Status of Women<
- Developing a Model for Other Communities

Evan D.J. Fraser wrote in the article Urban Ecology in Bangkok Thailand that although the project was initiated to serve the environmental needs of the city it quickly illustrated the positive side effects of urban agriculture.

"In many ways, the urban environment became a lens through which communities re-evaluated their own relationship with the city, the impact of urbanization in a global context, and how small groups can exert some control over the shape of their neighbourhoods."

Beijing, China

Beijing's increase in land area from 4,822 km^2 in 1956 to 16,808 km^2 in 1958 led to the increased adoption of peri-urban agriculture. Such "suburban agriculture" led to more than 70% of non-staple food in Beijing, mainly consisting of vegetables and milk, to be produced by the city itself in the 1960s and 1970s. Recently, with relative food security in China, periurban agriculture has led to improvements in the quality of the food available, as opposed to quantity. One of the more recent experiments in urban agriculture is the Modern Agricultural Science Demonstration Park in Xiaotangshan.

Shenzhen, China

Traditionally Chinese cities have been known to mix agricultural activities within the urban setting. Shenzhen, what was once a small farming community is now a fast growing metropolis due to the Chinese government designation as an open economic zone. Due to large and growing population in China, the government supports urban self-sufficiency in food production. Shenzhen's village structure, sustainable methods, and new agricultural advancements initiated by the government have been strategically configured to supply food for this growing city.

Village Structure, Methods, and New Advancements

- The city farms are located about 10 km from city centre in a two-tier system. The first tier approached from city centre produces perishable items. Located just outside these farms, hardier vegetables are grown such as potatoes, carrots, and onions. This system allows produce to be sold in city markets just a few short hours after picking.
- Another impressive method used within Chinese agriculture and aquaculture practice is the mulberry-dyke fish-pond system, which is a response to waste recycling and soil fertility. This system can be described as, "mulberry trees are grown to feed silkworms and the silkworm waste is fed to the fish in ponds. The fish also feed on waste from other animals, such as pigs, poultry, and buffalo. The animals in turn are given crops that have been fertilized by mud from the ponds. This is a sophisticated system as a continuous cycle of water, waste and food...with man built into the picture"
- As population grows and industry advances the city tries to incorporate potential agricultural growth by experimenting in new agricultural methods. The Fong Lau Chee Experimental Farm in Dongguan, Guangdong has worked with new agricultural advancements in lychee production. This farm was established with aspirations of producing large quantities and high quality lychees, by constantly monitoring sugar content, and their seeds. This research, conducted by local agricultural universities allows for new methods to be used with hopes of reaching the needs of city consumers.

Government Involvement

- However due to increased levels of economic growth and pollution some urban farms have become threatened. The government has been trying to step in and create new technological advancements within the agricultural field to sustain levels of urban agriculture.
- "The city plans to invest 8.82 billion yuan in 39 agricultural projects, including a safe agricultural base, an agricultural high-tech park, agricultural processing and distribution, forestry, eco-agricultural tourism, which will form an urban agriculture with typical Shenzhen characteristics" in conjunction with this programme the city is expected to expand the Buji Farm Produce Wholesale Market.
- According to the Municipal Bureau of Agriculture, Forestry and Fishery the city will invest 600 million yuan on farms located around the city, with hopes of the farms to provide "60 percent of the meat, vegetables and aquatic products in the Shenzhen market".

- There has also been an emerging trend of going green and organic as a response to pollution and pesticides used in farming practices. Vegetable suppliers are required to pass certain inspections held by the city's Agriculture Bureau before they can be sold as "green".

New York, NY

In New York City many low-income residents suffer from high rates of obesity and diabetes, limited sources of fresh produce and available, undeveloped land. The City and local nonprofit groups have been providing land, training and financial encouragement, but the impetus has in urban farming has really come from the farmers, who often volunteer when their regular work day is done.

Some urban gardeners have used empty lots to start community or urban garden. However, the soil must be tested for heavy contamination in city soil because of vehicle exhaust and remnants of old construction. However, studies have found that such ground can be cultivated as long as the pH is kept neutral. The City also has a composting programme, which is available to gardeners and farmers. One group, GreenThumb, provides free seedlings. Another programme, the City Farms project operated by the nonprofit Just Food, profits courses on growing and selling food.

BENEFITS OF URBAN AGRICULTURE

The benefits that UPA brings along to cities that implement this practice are numerous. The transformation of cities from only consumers of food to generators of agricultural products contributes to sustainability, improved health, and poverty alleviation.

- UPA assists to close the open loop system in urban areas characterized by the importation of food from rural zones and the exportation of waste to regions outside the city or town.
- Wastewater and organic solid waste can be transformed into resources for growing agriculture products: the former can be used for irrigation, the latter as fertilizer.
- Vacant urban areas can be used for agriculture production.
- Other natural resources can be conserved. The use of wastewater for irrigation increases the availability of freshwater for drinking and household consumption.
- UPA can help to preserve bioregional ecologies from being transformed into cropland.
- Urban agriculture saves energy (*e.g.* energy consumed in transporting food from rural to urban areas).
- Local production of food also allows savings in transportation costs, storage, and in product loss, what results in food cost reduction.

- UPA improves the quality of the urban environment through greening and thus, a reduction in pollution.

HEALTH IMPROVEMENT

- Urban agriculture also makes of the city a healthier place to live by improving the quality of the environment.
- UPA is a very efficient tool to fight against hunger and malnutrition since it facilitates the access to food by an impoverished sector of the urban population.

It is known that a large part of the people involved in urban agriculture is the urban poor. In developing countries, the majority of urban agricultural production is for self-consumption, with surpluses being sold in the market. According to the FAO (Food and Agriculture Organization of the United Nations), urban poor consumers spend between 60 and 80 percent of their income on food, making them very vulnerable to higher food prices.

- UPA provides food and creates savings in household expenditure on consumables, thus increasing the amount of income allocated to other uses.
- UPA surpluses can be sold in local markets, generating more income for the urban poor.

Community centers and gardens educate the community to see agriculture as an integral part of urban life. The Florida House Institute for Sustainable Development in Sarasota, Florida, serves as a public community and education centre in which innovators with sustainable, energy-saving ideas can implement and test them. Community centers like Florida House provide urban areas with a central location to learn about urban agriculture and to begin to integrate agriculture with the urban lifestyle. Other examples of community centers are Greensgrow Farm in Philadelphia and Northey Street City Farm in Brisbane, Australia. Greensgrow uses an abandoned site as an urban farm to teach the community how food is grown and how to grow their own food. Northey Street City Farm hosts weekly community activities to educate and involve local residents in agricultural practices. Urban farms also are a proven effective educational tool to teach kids about healthy eating and meaningful physical activity. An example of educational urban agriculture is Full Circle Farm, an 11-acre farm located on a middle school campus in the heart of Silicon Valley. Like other educational agriculture centers, Full Circle Farm's acreage is used as a "living campus" where students get real-world, hands-on agriculture experiences that cultivate both healthy habits and environmental leadership.

DIFFICULTIES OF URBAN AGRICULTURE

- The use of waste water for irrigation without careful treatment and monitoring can result in the spread of diseases among the population.

- Cultivation on contaminated land represents also a health hazard for the consumers.
- The practice of cultivating along roadsides facilitates the distribution of products to local markets, but it is at the same time a risky practice since it exposes food to car pollution.
- Agriculture and urbanization are considered to be incompatible activities, competing for the access and use of limited land. In reality, in urban areas there is important available space for agriculture use such as public and private vacant lots, and areas not suited for built-up uses (steep slopes and flood plains).
- Legal restrictions and economic impediments to accessing land and resources (such as reasonably priced water) are among the most common problems confronted by urban agriculture.
- Lack of security of tenure also acts as a preventive for farming due to the uncertainty in the use length of the land.
- Urban agriculture has been criticized by those who believe that industrial farm production can produce food at larger volumes more efficiently.
- A major argument is whether urban farming alone-that is farming very intensively on small land areas-could replace land extensive production in rural areas which produce the bulk of our food products. Yet, hunger persists in both urban and rural areas, despite a subsidized industrial agriculture. The degree to which urban agriculture can address these food needs systemically is undetermined, though there are indications in some communities it is an important source of food.
- Other opponents argue that localized food production and the introduction of common resources and common lands into the urban areas would produce a tragedy of the commons. Though, as referenced earlier, many urban farms and community gardens are managed privately or through other civil society organizations.

NEOLITHIC REVOLUTION

The Neolithic Revolution was the first agricultural revolution—the transition from hunting and gathering communities and bands, to agriculture and settlement (settlement is currently being questioned). Archaeological data indicate that various forms of domestication of plants and animals arose independently in at least 7-8 separate locales worldwide, with the earliest known developments taking place in the Middle East around 10,000 BC (BCE) or earlier.

However, the Neolithic Revolution involved far more than the adoption of a limited set of food-producing techniques. During the next millennia it would transform the small, mobile and fairly egalitarian groups of hunter-gatherers that had hitherto dominated human history, into sedentary societies based in built-up villages and towns, which radically modified their natural environment by means of specialized cultivation and storage technologies (*e.g.* irrigation) that allowed extensive surplus production. These developments provided the basis for high population densities, complex labor diversification, trading economies, centralized administrations and political structures, hiearchical ideologies and depersonalized systems of knowledge (*e.g.* property regimes and writing). The first full-blown manifestation of the entire Neolithic complex is seen in the Middle Eastern Sumerian cities (ca. 5,300 BC), whose emergence also inaugurates the end of the prehistoric Neolithic and the beginning of historical time.

The relationship of the above-mentioned Neolithic characteristics to the onset of agriculture, their sequence of emergence and empirical relation to each other at various Neolithic sites remains the subject of academic debate, and seems to vary from place to place, rather than being the outcome of universal laws of social evolution.

AGRICULTURAL TRANSITION

The term *Neolithic Revolution* was coined in the 1920s by Vere Gordon Childe to describe the first in a series of agricultural revolutions in Middle Eastern history. The period is described as a "revolution" to denote its importance, and the great significance and degree of change affecting the communities in which new agricultural practices were gradually adopted and refined.

The beginning of this process in different regions has been dated from perhaps 10,000 years ago in Melanesia to 2,500 BC in Subsaharan Africa, with some considering the developments of 9000-7000 BC in the Fertile Crescent to be the most important. This transition everywhere seems associated with a change from a largely nomadic hunter-gatherer way of life to a more settled, agrarian-based one, with the inception of the domestication of various plant and animal species-depending upon which species were locally available, and probably also influenced by local culture.

There are several competing (but not mutually exclusive) theories as to the factors which drove populations to take up agriculture. The most prominent of these are:

- The Oasis Theory, originally proposed by Raphael Pumpelly in 1908, popularized by Vere Gordon Childe in 1928 and summarised in Childe's book *Man Makes Himself.*, which maintains that as the climate got drier, communities contracted to oases where they were forced into close association with animals, which were then domesticated together with planting of seeds. It has little support now because climate data for the time does not support the theory.

- The Hilly Flanks hypothesis, proposed by Robert Braidwood in 1948, suggests that agriculture began in the hilly flanks of the Taurus and Zagros mountains, where the climate was not drier as Childe had believed, and fertile land supported a variety of plants and animals amenable to domestication.
- The Feasting model by Brian Hayden suggests that agriculture was driven by ostentatious displays of power, such as giving feasts, to exert dominance. This required assembling large quantities of food which drove agricultural technology.
- The Demographic theories proposed by Carl Sauer and adapted by Lewis Binford and Kent Flannery posit an increasingly sedentary population which expanded up to the carrying capacity of the local environment and required more food than could be gathered. Various social and economic factors helped drive the need for food.
- The evolutionary/intentionality theory, developed by David Rindos and others, views agriculture as an evolutionary adaptation of plants and humans. Starting with domestication by protection of wild plants, it led to specialization of location and then full-fledged domestication.

In contrast to the Paleolithic, in which more than one hominid species existed, only one (Homo sapiens) reached the Neolithic.

DOMESTICATION OF PLANTS

Once agriculture started gaining momentum, cereal grasses (beginning with emmer, einkorn and barley), and not simply those that would favour greater caloric returns through larger seeds, were selectively bred. Plants that possessed traits such as small seeds or bitter taste would have been seen as undesirable. Plants that rapidly shed their seeds on maturity tended not to be gathered at harvest, thus not stored and not seeded the following season; years of harvesting selected for strains that retained their edible seeds longer. Several plant species, the "pioneer crops" or Neolithic founder crops, were the earliest plants successfully manipulated by humans. Some of these pioneering attempts failed at first and crops were abandoned, sometimes to be taken up again and successfully domesticated thousands of years later: rye, tried and abandoned in Neolithic Anatolia, made its way to Europe as weed seeds and was successfully domesticated in Europe, thousands of years after the earliest agriculture. Wild lentils present a different challenge that needed to be overcome: most of the wild seeds do not germinate in the first year; the first evidence of lentil domestication, breaking dormancy in their first year, was found in the early Neolithic at Jerf el-Ahmar, (in modern Syria), and quickly spread south to the Netiv HaGdud site in the Jordan Valley. This process of domestication allowed the founder crops to adapt and eventually become larger, more easily harvested, more dependable in storage and more useful to the human population. Figs, barley and, most likely, oats were cultivated in the Jordan Valley, represented by the early Neolithic site of Gilgal, where in 2006 archaeologists found caches of seeds of each in quantities too large to be accounted for even by intensive gathering, at strata dateable c. 11,000 years ago. Some of the plants tried and then abandoned during the Neolithic period in the Ancient Near East, at sites like Gilgal, were later successfully domesticated in other parts of the world.

Once early farmers perfected their agricultural techniques, their crops would yield surpluses which needed storage. Most hunter gatherers could not easily store food for long due to their migratory lifestyle, whereas those with a sedentary dwelling could store their surplus grain. Eventually granaries were developed that allowed villages to store their seeds for longer periods of time. So with more food, the population expanded and communities developed specialized workers and more advanced tools.

The process was not as linear as was once thought, but a more complicated effort, which was undertaken by different human populations in different regions in many different ways.

AGRICULTURE IN ASIA

The Neolithic Revolution is believed to have become widespread in southwest Asia around 8000 BC–7000 BC, though earlier individual sites have been identified. Although archaeological evidence provides scant evidence as to which of the genders performed what task in Neolithic cultures, by comparison with historical and contemporary hunter-gatherer communities it is generally supposed that hunting was typically performed by the men, whereas women had a more significant role in the gathering. By extension, it may be theorised that women were largely responsible for the observations and initial activities which began the Neolithic Revolution, insofar as the gradual selection and refinement of edible plant species was concerned.

The precise nature of these initial observations and (later) purposeful activities which would give rise to the changes in subsistence methods brought about by the Neolithic Revolution are not known; specific evidence is lacking. However, several reasonable speculations have been put forward; for example, it might be expected that the common practice of discarding food refuse in middens would result in the regrowth of plants from the discarded seeds in the (fertilizer-enriched) soils. In all likelihood, there were a number of factors which contributed to the early onset of agriculture in Neolithic human societies.

AGRICULTURE IN THE FERTILE CRESCENT

Generalised agriculture apparently first arose in the Fertile Crescent because of many factors. The Mediterranean climate has a long dry season with a short period of rain, which made it suitable for small plants with large seeds, like wheat and barley. These were the most suitable for domestication because of the ease of harvest and storage and the wide availability. In addition, the domesticated plants had especially high protein content. The Fertile Crescent had a large area of varied geographical settings and altitudes. The variety given made agriculture more profitable for former hunter-gatherers. Other areas with a similar climate were less suitable for agriculture because of the lack of geographic variation within the region and the lack of availability of plants for domestication.

AGRICULTURE IN AFRICA

The Revolution developed independently in different parts of the world, not just in the Fertile Crescent. On the African continent, three areas have been identified as independently developing agriculture: the Ethiopian highlands, the Sahel and West Africa.

The most famous crop domesticated in the Ethiopian highlands is coffee. In addition, Khat, Ensete, Noog, teff and finger millet were also domesticated in the Ethiopian highlands. Crops domesticated in the Sahel region include sorghum and pearl millet. The Kola nut, extracts from which became an ingredient in Coca Cola, was first domesticated in West Africa. Other crops domesticated in West Africa include African rice, African yams and the oil palm.

A number of crops that have been cultivated in Africa for millennia came after their domestication elsewhere. Agriculture in the Nile River Valley developed from crops domesticated in the Fertile Crescent. Bananas and plantains which were first domesticated in Southeast Asia, most likely Papua New Guinea, were re-domesticated in Africa possibly as early as 5,000 years ago. Asian yams and taro were also cultivated in Africa.

Prof. Fred Wendorf and Dr. Romuald Schild, of the Department of Anthropology at Southern Methodist University, originally thought to have found evidence of early agriculture in Upper Paleolithic times at Wadi Kubbaniya, on the Kom Ombos plateau, of Egypt, including a mortar and pestle, grinding stones, several harvesting implements and charred wheat and barley grains — which may have been introduced from outside the region. AMS dating since their first reports has invalidated their hypothesis. Many such grinding stones are found with the early Egyptian Sebilian and Mechian cultures and evidence has been found of a neolithic domesticated crop-based economy dating around 5000 BC. Smith writes: "With the benefit of hindsight we can now see that many Late Paleolithic peoples in the Old World were poised on the brink of plant cultivation and animal husbandry as an alternative to the hunter-gatherer's way of life". Unlike the Middle East, this evidence appears as a "false dawn" to agriculture, as the sites were later abandoned, and permanent farming then was delayed until 4,500 BC with the Tasian and Badarian cultures and the arrival of crops and animals from the Near East.

AGRICULTURE IN THE AMERICAS

Corn, beans and squash were domesticated in Mesoamerica around 3500 BCE. Potatoes and manioc were domesticated in South America. In what is now the eastern United States, Native Americans domesticated sunflower, sumpweed and goosefoot around 2500 BCE.

DOMESTICATION OF ANIMALS

When hunter-gathering began to be replaced by sedentary food production it became more profitable to keep animals close at hand. Therefore, it became necessary to bring animals permanently to their settlements, although in many cases there was a distinction between relatively sedentary farmers and nomadic herders. The animals' size, temperament, diet, mating patterns, and life span were factors in the desire and success in domesticating animals. Animals that provided milk, such as cows and goats, offered a source of protein that was renewable and therefore quite valuable. The animal's ability as a worker (for example ploughing or towing), as well as a food source, also had to be taken into account. Besides being a direct source of food, certain animals could provide leather, wool, hides, and fertilizer. Some of the earliest domesticated animals included dogs (about 15,000 years ago), sheep, goats, cows, and pigs.

DOMESTICATION OF ANIMALS IN THE MIDDLE EAST

The Middle East served as the source for many animals that could be domesticated, such as goats and pigs. This area was also the first region to domesticate the Dromedary Camel. The presence of these animals gave the region a large advantage in cultural and economic development. As the climate in the Middle East changed, and became drier, many of the farmers were forced to leave, taking their domesticated animals with them. It was this massive emigration from the Middle East that would later help distribute these animals to the rest of Afroeurasia.

This emigration was mainly on an east-west axis of similar climates, as crops usually have a narrow optimal climatic range outside of which they cannot grow for reasons of light or rain changes. For instance, wheat does not normally grow in tropical climates, just like tropical crops such as bananas do not grow in colder climates.

Some authors like Jared Diamond postulated that this East-West axis is the main reason why plant and animal domestication spread so quickly from the Fertile Crescent to the rest of Eurasia and North Africa, while it did not reach through the North-South axis of Africa to reach the Mediterranean climates of South Africa, where temperate crops were successfully imported by ships in the last 500 years. The African Zebu is a separate breed of cattle that was better suited to the hotter climates of central Africa than the fertile-crescent domesticated bovines. North and South America were similarly separated by the narrow tropical Isthmus of Panama, that prevented the andes llama to be exported to the Mexican plateau.

CAUSES OF THE NEOLITHIC REVOLUTION

Jack Harlan, examining the causes for the Neolithic Revolution, suggests 6 principal reasons which can be summarized to 3 principal categories:

1. Domestication for religious reasons
2. Domestication by crowding and as a consequence of stress
3. Domestication resulting from discovery, based upon the perceptions of food gatherers

With regard to the first explanation, Ian Hodder, who directs the excavations at Çatalhöyük, has said that the earliest settled communities, and the Neolithic revolution they represent, actually *preceded* the development of agriculture. He has been developing the ideas first expressed by Jacques Cauvin, the excavator of the Natufian settlement at Mureybet in northern Syria. Hodder believes that the Neolithic revolution was the result of a revolutionary change in the human psychology, a "revolution of symbols" which led to new beliefs about the world and shared community rituals embodied in corpulent female figurines and the methodical assembly of aurochs horns. An alternative explanation for the origin of agriculture has been advanced by anthropologist Mark Nathan Cohen. Cohen believes that following the widespread extinctions of large mammals in the late Palaeolithic, the human population had expanded to the limits of the available territory and a population explosion led to a food crisis. Agriculture was the only way in which it was possible to support the increasing population on the available area of land. This view has come under criticism due to the obvious problem of how a population explosion would occur without already having a surplus of food.

Food gatherers (not the hunters) caring for children, keeping the fires alive, and foraging near the base camp, led the way in developing language and culture, in knowledge of plants, and increasingly semi-domesticated animals who travelled with the nomads from camp to camp.

CONSEQUENCES OF THE NEOLITHIC REVOLUTION

SOCIAL CHANGE

It is often argued that agriculture gave humans more control over their food supply, but this has been disputed by the finding that nutritional standards of Neolithic populations were generally inferior to that of hunter gatherers, and life expectancy may in fact have been shorter, in part due to diseases. Actually, by reducing the necessity for the carrying of children, Neolithic societies had a major impact upon the spacing of children (carrying more than one child at a time is impossible for hunter-gatherers, which leads to children being spaced four or more years apart). This increase in the birth rate was required to offset increases in death rates and required settled occupation of territory and encouraged larger social groups. These sedentary groups were able to reproduce at a faster rate due to the possibilities of sharing the raising of children in such societies. The children accounted for a denser population, and encouraged the introduction of specialization by providing diverse forms of new labor. The development of larger societies seemed to have led to the development of different means of decision making and to governmental organization. Food surpluses made possible the development of a social elite who were not otherwise engaged in agriculture, industry or commerce, but dominated their communities by other means and monopolized decision-making.

The Neolithic Revolution had many results socially and economically. Socially people could have more children and social classes emerged. Economically a surplus of food meant that people could also specialize in different things. The Neolithic Revolution was clearly the start of our society as we know it today. About twelve thousand to six thousand years ago humans started to notice changes of season and when and how to grow food. The slow process of learning how to use the land brought about many changes in the world socially. The ability to know where food was, how to grow it and when you could eat it gave humans enough security to start settling and replace their nomadic hunter gatherer ways with agriculture and villages. As humans grew better and better at farming, there became a surplus of food. This resulted in an explosion of population as demonstrated in this chart:

Year	*Human Population*
3000 B.C.	14 million
2000 B.C.	27 million
1000 B.C.	50 million
500 B.C.	100 million

Social evolution of humans

Period years ago	*Society type*	*Number of individuals*
100,000-10,000	Bands	10s-100s
10,000-5000	tribes	100s-1,000s
5,000-3,000	Chiefdoms	1,000s-10,000s
3,000-1,000	States	10,000s-100,000s
1,000-Present	Empires	100,000-1,000,000s

• Traditions and Encounters textbook, page 24

The population nearly doubles every millennium as humans get better with agriculture and then from 1000 B.C. to 500 B.C. (only 500 years as compared to the 1000 years of the previous set) the population doubles in half the time. Agriculture also changed economics entirely. When humans were still nomadic hunter gatherers they had no time during the day to teach their children things other than how to hunt and gather. Finally, when the people could settle in villages and have a steady source of food do they really have time to take up special labor, "The concentration of large numbers of people in villages encouraged specialization of labor".

The best known Neolithic settlement that demonstrates this is Çatal Hüyük. At this Neolithic village archeologists have found "pots, baskets, textiles, leather, stone and metal, wood carvings, carpets, beads, and jewelry...". Pottery, metallurgy, and textile production were the main three industries that started at that time. After people started to settle and specialize and have more children the egalitarian society of the hunter gatherers also changed. Social classes emerged as familial wealth gets passed from one generation to the next. At Çatal Hüyük archeologists have found differences in interior decoration of one house to another, which clearly shows one family's wealth as compared to another's.

SUBSEQUENT REVOLUTIONS

Andrew Sherratt has argued that following upon the Neolithic Revolution was a second phase of discovery that he refers to as the secondary products revolution. Animals, it appears were first domesticated purely as a source of meat. The Secondary Products Revolution occurred when it was recognised that animals also provided a number of other useful products. These included:

- hides and skins (from no domesticated animals)
- manure for soil conditioning (from all domesticated animals)
- wool (from sheep, llamas, alpacas, and Angora goats)
- milk (from goats, cattle, yaks, sheep, horses and camels)
- traction (from oxen, onagers, donkeys, horses and camels)

Sherratt argues that this phase in agricultural development enabled humans to make use of the energy possibilities of their animals in new ways, and permitted permanent intensive subsistence farming and crop production, and the opening up heavier soils for farming. It also made possible nomadic pastoralism in semi arid areas, along the margins of deserts, and eventually led to the domestication of both the dromedary and bactrian camel. Overgrazing of these areas, particularly by herds of goats, greatly extended the areal extent of deserts. Living in one spot would have more easily permitted the accrual of personal possessions and an attachment to certain areas of land. From such a position, it is argued, prehistoric people were able to stockpile food to survive lean times and trade unwanted surpluses with others. Once trade and a secure food supply were established, populations could grow, and society would have diversified into food producers and artisans, who could afford to develop their trade by virtue of the free time they enjoyed because of a surplus of food. The artisans, in turn, were able to develop technology such as metal weapons. Such relative complexity would have required some form of social organisation to work efficiently and so it is likely that populations which had such organisation, perhaps such as that provided by religion were better prepared and more successful. In addition, the denser populations could form and support legions of professional soldiers. Also, during this time property ownership became increasingly important to all people. Ultimately, Childe argued that this growing social complexity, all rooted in the original decision to settle, led to a second Urban Revolution in which the first cities were built.

DISEASE

Throughout the development of sedentary societies, disease spread more rapidly than it had during the time in which hunter-gatherer societies existed. Inadequate sanitary practices and the domestication of animals may explain the rise in deaths and sickness during the Neolithic Revolution from disease, as diseases jumped from the animal to the human population. Some examples of diseases spread from animals to humans are influenza, smallpox, and measles. In concordance with a process of natural selection, the humans who first domesticated the big mammals quickly built up immunities to the diseases as within each generation the individuals with better immunities had better chances of survival.

In their approximately 10,000 years of shared proximity with animals, Eurasians and Africans became more resistant to those diseases compared with the indigenous populations encountered outside Eurasia and Africa. For instance, the population of most Caribbean and several Pacific Islands have been completely wiped out by diseases. According to the Population history of American indigenous peoples, 90% of the population of certain regions of North and South America were wiped out long before direct contact with Europeans. Some cultures like the Inca Empire did have one big mammal domesticated, the Llama, but the Inca did not drink its milk or live in a closed space with their herds, hence limiting the risk of contagion.

The causal link between the type or lack of agricultural development, disease and colonisation is not supported by colonization in other parts of the world. Disease increased after the establishment of British Colonial rule in Africa and India despite the areas having diseases that Europeans had no natural immunity to. In India agriculture developed during the Neolithic period with a wide range of animals domesticated. During colonial rule an estimated 23 million people died from cholera between 1865 and 1949, and millions more died from plague, malaria, influenza and tuberculosis.

In Africa European colonisation was accompanied by great epidemics, including malaria and sleeping sickness and despite parts of colonised Africa having little or no agriculture Europeans were more susceptible than these Africans. The increase of disease has been attributed to increased mobility of people, increased population density, urbanisation, environmental deterioration and irrigation schemes that helped to spread malaria rather than the development of agriculture.

TECHNOLOGY

In his book *Guns, Germs, and Steel,* Jared Diamond argues that Europeans and East Asians benefited from an advantageous geographical location which afforded them a head start in the Neolithic Revolution. Both shared the temperate climate ideal for the first agricultural settings, both were near a number of easily domesticable plant and animal species, and both were safer from attacks of other people than civilizations in the middle part of the Eurasian continent.

Being among the first to adopt agriculture and sedentary lifestyles, and neighboring other early agricultural societies with whom they could compete and trade, both Europeans and East Asians were also among the first to benefit from technologies such as firearms and steel swords. In addition, they developed resistances to infectious disease, such as smallpox, due to their close relationship with domesticated animals. Groups of people who had not lived in proximity with other large mammals, such as the Australian Aborigines and American indigenous peoples were more vulnerable to infection and largely wiped out by diseases.

During and after the Age of Discovery, European explorers, such as the Spanish conquistadors, encountered other groups of people who had never or only recently adopted agriculture, such as in the Pacific Islands, or lacked domesticated big mammals such as the highlands people of Papua New Guinea.

MUSLIM AGRICULTURAL REVOLUTION

The Islamic Golden Age from the 8th century to the 13th century witnessed a fundamental transformation in agriculture known as the Arab Agricultural Revolution, Medieval Green Revolution, or Muslim Agricultural Revolution. The global economy established by Arab and other Muslim traders across the Old World, enabled the diffusion of many crops and farming techniques among different parts of the Islamic world, as well as the adaptation of crops and techniques from and to regions beyond the Islamic world. Crops from Africa such as sorghum, crops from China such as citrus fruits, and numerous crops from India such as mangos, rice, cotton and sugar cane, were distributed throughout Islamic lands, which previously had not grown these crops. Some writers have referred to the diffusion of numerous crops during this period as the Globalization of crops. These introductions, along with an increased mechanization of agriculture, led to major changes in economy, population distribution, vegetation cover, agricultural production and income, population levels, urban growth, the distribution of the labour force, linked industries, cooking, diet and clothing in the Islamic world.

AGE OF DISCOVERY

The earliest forms of globalization began emerging during the Islamic Empire and the Islamic Golden Age, when the knowledge, trade and economies from many previously isolated regions and civilizations began integrating due to contacts with Muslim explorers, sailors, scholars, traders, and travelers. Some have called this period the "Pax Islamica" or "Afro-Asiatic age of discovery", in reference to the Jewish (known as Radhanites), Sogdian and Muslim Southwest Asian and North African traders and explorers who travelled most of the Old World, and established an early global economy across most of Asia and Africa and much of Europe, with their trade networks extending from the Atlantic Ocean and Mediterranean Sea in the west to the Indian Ocean and China Sea in the east. This helped establish the Islamic Empire (including the Rashidun, Umayyad, Abbasid and Fatimid caliphates) as the world's leading extensive economic power throughout the 7th-13th centuries. Several contemporary medieval Arabic reports also suggest that Muslim explorers from al-Andalus and the Maghreb may have travelled in expeditions across the Atlantic Ocean, possibly even to the Americas, between the 9th and 14th centuries.

AGRICULTURAL INNOVATIONS

Muslims widely practiced cash cropping and the modern crop rotation system where land was cropped four or more times in a two-year period. Winter crops were followed by summer ones, and in some cases there was in between. In areas where plants of shorter growing season were used, such as spinach and eggplants, the land could be cropped three or more times a year. In parts of Yemen, wheat yielded two harvests a year on the same land, as did rice in Iraq. Muslims developed a scientific approach based on three major elements; sophisticated systems of crop rotation, highly developed irrigation techniques, and the introduction of a large variety of crops which were studied and catalogued according to the season, type of land and amount of water they require. Numerous encyclopaedias on farming and botany were produced, with highly accurate precision and details.

ADVANCED AGRICULTURAL SYSTEMS

As early as the 9th century, an essentially modern agricultural system became central to economic life and organization in the Arab caliphates, replacing the largely export driven Roman model. Cities of the Near East, North Africa, and Moorish Spain were supported by elaborate agricultural systems which included extensive irrigation based on knowledge of hydraulic and hydrostatic principles, some of which were continued from Roman times.

In later centuries, Persian Muslims began to function as a conduit, transmitting cultural elements, including advanced agricultural techniques, into Turkic lands and western India. The Muslims introduced what was to become an agricultural revolution based on four key areas:

- Development of a sophisticated system of irrigation using machines such as norias, water mills, water raising machines, dams and reservoirs. With such technology they managed to greatly expand the exploitable land area.
- The adoption of a scientific approach to farming enabled them to improve farming techniques derived from the collection and collation of relevant information throughout the whole of the known world. Farming manuals were produced in every corner of the Muslim world detailing where, when and how to plant and grow various crops. Advanced scientific techniques allowed leaders like Ibn al-Baitar to introduce new crops and breeds and strains of livestock into areas where they were previously unknown.
- Incentives based on a new approach to land ownership and labourers' rights, combining the recognition of private ownership and the rewarding of cultivators with a harvest share commensurate with their efforts. Their counterparts in Europe struggled under a feudal system in which they were almost slaves (serfs) with little hope of improving their lot by hard work.
- The introduction of new crops transforming private farming into a new global industry exported everywhere, including Europe, where farming was mostly restricted to wheat strains obtained much earlier via central Asia. Spain received what she in turn transmitted to the rest of Europe; many agricultural and fruit-growing processes, together with many new plants, fruit and vegetables. These new crops included sugar cane, rice, citrus fruit, apricots, cotton, artichokes, aubergines, and saffron. Others, previously known, were further developed. Muslims also brought to that country lemons, oranges, cotton, almonds, figs and sub-tropical crops such as bananas and sugar cane. Several were later exported from Spanish coastal areas to the Spanish colonies in the New World. Also transmitted via Muslim influence, a silk industry flourished, flax was cultivated and linen exported, and esparto grass, which grew wild in the more arid parts, was collected and turned into various articles.

ECONOMIC AND SOCIAL REFORMS

The caliphate understood that incentives were needed to increase productivity and wealth, thus enhancing tax revenues. Hence, they introduced a social transformation through the changed ownership of land, where any individual of any gender or any ethnic or religious background had the right to buy, sell, mortgage and inherit land for farming or any other purposes. They also introduced the signing of a contract for every major financial transaction concerning agriculture, industry, commerce, and employment. Copies of the contract was usually kept by both parties involved. The two types of economic systems that prompted agricultural development in the Islamic world were either politically-driven, by the conscious decisions of the central authority to develop under-exploited lands; or market-driven, involving the spread of advice, education, and free seeds, and the introduction of high value crops or animals to areas where they were previously unknown.

These led to increased subsistence, a high level of economic security that ensured wealth for all citizens, and a higher quality of life due to the introduction of artichokes, spinach, aubergines, carrots, sugar cane, and various exotic plants; vegetables being available all year round without the need to dry them for winter; citrus and olive plantations becoming a common sight, market gardens and orchards springing up in every Muslim city; intense cropping and the technique of intensive irrigation agriculture with land fertility replacement; a major increase in animal husbandry; higher quality of wool and other clothing materials; and the introduction of selective breeding of animals from different parts of the Old World resulting in improved horse stocks and the best load-carrying camels.

The average life expectancy in the lands under Islamic rule also experienced an increase, due to the Agricultural Revolution as well as improved medical care. In contrast to the average lifespan in the ancient Greco-Roman world (22-28 years), the average lifespan in the early Islamic Caliphate was more than 35 years. The average lifespans of the Islamic scholarly class in particular was much higher: 84.3 years in 10th-11th century Iraq and Persia, 72.8 years in the 11th century Middle East, 69-75 years in 11th century Islamic Spain, 75 years in 12th century Persia, and 59-72 years in 13th century Persia. The Islamic Empire also experienced a growth in literacy, having the highest literacy rate of the Middle Ages, comparable to Athens' literacy in Classical Antiquity but on a larger scale.

SUGAR INDUSTRY

During the Muslim Agricultural Revolution, sugar production was refined and transformed into a large-scale industry by the Arabs. The Arabs and Berbers diffused sugar throughout the Arab Empire from the 8th century.

OTHER INNOVATIONS

Many other agricultural innovations were introduced by Muslim farmers and engineers, such as new forms of land tenure, improvements in irrigation, a variety of sophisticated irrigation methods, the introduction of fertilizers and widespread artificial irrigation systems, the development of gravity-flow irrigation systems from rivers and springs, the first uses of noria and chain pumps for irrigation purposes, the establishment of the sugar cane industry in the Mediterranean and experimentation in sugar cultivation, numerous advances in industrial milling and water-raising machines, and many other improvements and innovations.

AGRICULTURAL SCIENCES

During the Muslim Agricultural Revolution, Muslim scientists laid the foundations of agricultural science, which included significant advances in the fields of agronomy, astronomy, botany, earth science, environmental philosophy, and environmental science. In particular, the experimental scientific method was introduced into the field in the 13th century by the Andalusian-Arab botanist Abu al-Abbas al-Nabati, the teacher of Ibn al-Baitar. Al-Nabati introduced empirical techniques in the testing, description and identification of numerous materia medica, and he separated unverified reports from those supported by actual tests and observations.

The earliest known work dedicated to the study of agriculture was Ibn Wahshiyya's *Nabatean Agriculture,* which also dealt with the related field of botany and was also an early cookbook. The early Arab lexicographs were the first known works to separate the two disciplines of agriculture and botany, though both were considered part of the medical sciences due to agriculture's primary role being to feed and botany's primary role being to heal. The agricultural sciences were known by the Arabic term *filaha,* which had a dual-meaning, to both care for the Earth and to take care of plants. Many of the early Islamic authors on botany were often philologists, due to their role in the translation of ancient scientific texts. This was also the case with early Arabic zoology, like with al-Jahiz for example.

Al-Asma'i was the earliest known Arab biologist, botanist and zoologist; his works include the *Book of Distinction, Book of the Wild Animals, Book of the Horse,* and *Book of the Sheep.*

AGRONOMY

Muslim agriculturists demonstrated advanced agronomic, agrotechnical and economic knowledge in areas such as meteorology, climatology, hydrology, soil occupation, and the economy and management of agricultural enterprises. They also demonstrated agricultural knowledge in areas such as pedology, agricultural ecology, irrigation, preparation of soil, planting, spreading of manure, killing herbs, sowing, cutting trees, grafting, pruning vine], prophylaxis, phytotherapy, the care and improvement of microbiological cultures and plants, and the harvest and storage of crops.

The *Nabatean Agriculture* then goes on to discuss a number of other complex issues on agriculture, including the management of an agricultural enterprise and the duties of the owner regarding his enterprise and workers;

the official (*wakil*) in charge of the management of the enterprise, his obligation towards the farmer]s, and applying the instructions he receives from his boss; the weather forecasting of atmospheric changes and signs from the planetary astral alterations; signs of rain based on observation of the lunar phases, nature of thunder and lightning, direction of sunrise, behaviour of certain plants and animals, and weather forecasts based on the movement of winds; the recognition of plant tissue cultures which succeed in certain years; a list of work to be done in each month of year; the position of the moon relative to the Earth; the required knowledge of a farmer and the owner of an agricultural enterprise; pollenized air and winds; and formation of winds and vapours.

Other agricultural topics discussed in the *Nabatean Agriculture* include the causes of the corruption of plants and of torrential rain; the nature of soils and their different flavours; the manure; how to get rid of bad herbs and how to cut plants which need to be cut; and a number of other agricultural topics.

In 12th century al-Andalus, Ibn al-'Awwam al-Ishbili wrote the *Kitab al-Filaha* which synthesized his own agricultural knowledge with that of the *Nabatean Agriculture* and his other Arabic predecessors. This work also described 585 microbiological cultures, 55 of which concern fruit trees. This work was influential in Europe after it was translated into Spanish by Banqueri in Madrid in 1801 and into French by Clement-Mullet in Paris in 1864.

ASTRONOMY AND METEOROLOGY

Another innovation during this period was the application of astronomy to agriculture and botany. As weather forecasting predictions and the measurement of time and the onset of seasons became more precise and reliable, farmers became informed of these advances and often employed them in agriculture. They also benefited from the compilation of calendars with information on when to plant each type of crop, when to graft trees, when and how to fertilize crops, when to harvest, and what to eat and what to avoid at each time of year. These advances made it possible for farmers to plan the growth of each of their crops for specific markets and at specific times of the year. Parts of al-Dinawari's *Book of Plants* deal with the applications of astronomy and meteorology to agriculture. It describes the astronomical and meteorological character of the sky, the planets and constellations, the sun and moon, the lunar phases indicating seasons and rain, the *anwa* (heavenly bodies of rain), and atmospheric phenomena such as winds, thunder, lightning, snow, floods, valleys, rivers, lakes, wells and other sources of water.

BOTANY

Muslims developed a scientific approach to botany and agriculture based on three major elements: sophisticated systems of crop rotation, highly developed irrigation techniques, and the introduction of a large variety of crops which were studied and catalogued according to the season, type of land and amount of water they require. Numerous encyclopaedias on botany were produced, with highly accurate precision and details.

The 9th century botanist al-Dinawari is considered the founder of Arabic botany. He wrote a botanical encyclopedia entitled *Kitab al-Nabat* (*Book of Plants*), which consisted of six volumes. Only the third and fifth volumes have survived, though the sixth volume has partly been reconstructed based on citations from later works. In the surviving portions of his works, 637 plants are described from the letters *sin* to *ya*. He also discusses plant evolution from its birth to its death, describing the phases of plant growth and the production of flowers and fruit.

In the early 13th century, Ibn al-Baitar published the *Kitab al-Jami fi al-Adwiya al-Mufrada,* which is considered one of the greatest botanical compilations and pharmaceutical encyclopedias, and was a botanical authority for centuries. It contains details on at least 1,400 different plants, foods, and drugs, 300 of which were his own original discoveries. The *Kitab al-Jami fi al-Adwiya al-Mufrada* was also influential in Europe after it was translated into Latin in 1758, where it was being used up until the early 19th century.

EARTH SCIENCE

Muslim scientists made a number of contributions to the earth sciences. Alkindus introduced experimentation into the Earth sciences. Parts of al-Dinawari's *Book of Plants* deals with the Earth sciences in the context of agriculture. He considers the Earth, stone and sands, and describes different types of ground, indicating which types are more convenient for plants and the qualities and properties of good ground.

Biruni is considered the father of geodesy for his important contributions to the field, along with his significant contributions to geography and geology. Among his writings on geology, Biruni wrote the following on the geology of India:

"But if you see the soil of India with your own eyes and meditate on its nature, if you consider the rounded stones found in earth however deeply you dig, stones that are huge near the mountains and where the rivers have a violent current: stones that are of smaller size at a greater distance from the mountains and where the streams flow more slowly: stones that appear pulverised in the shape of sand where the streams begin to stagnate near their mouths and near the sea-if you consider all this you can scarcely help thinking that India was once a sea, which by degrees has been filled up by the alluvium of the streams."

John J. O'Connor and Edmund F. Robertson write in the *MacTutor History of Mathematics archive*:

"Important contributions to geodesy and geography were also made by al-Biruni. He introduced techniques to measure the earth and distances on it using triangulation. He found the radius of the earth to be 6339.6 km, a value not obtained in the West until the 16th century. His Masudic canon contains a table giving the coordinates of six hundred places, almost all of which he had direct knowledge."

In geology, Avicenna hypothesized on two causes of mountains in *The Book of Healing*. In cartography, the Piri Reis map drawn by the Ottoman cartographer Piri Reis in 1513, was one of the earliest world maps to include the Americas, and perhaps the first to include Antarctica. His map of the world was considered the most accurate in the 16th century.

ENVIRONMENTAL PHILOSOPHY

Perhaps due to resource scarcity in most Islamic nations, there was an emphasis on limited (and some claim also sustainable) use of natural capital, *i.e.* producing land.

Traditions of haram and hima and early urban planning were expressions of strong social obligations to stay within carrying capacity and to preserve the natural environment as an obligation of khalifa or "stewardship".

Muhammad is considered a pioneer of environmentalism for his teachings on environmental preservation. His hadiths on agriculture and environmental philosophy were compiled in the "Book of Agriculture" of the *Sahih Bukhari*, which included the following saying:

"There is none amongst the believers who plants a tree, or sows a seed, and then a bird, or a person, or an animal eats thereof, but it is regarded as having given a charitable gift [for which there is great recompense]."

Several such statements concerning the environment are also found in the Qur'an, such as the following:

"And there is no animal in the earth nor bird that flies with its two wings, but that they are communities like yourselves."

ENVIRONMENTAL SCIENCE

The earliest known treatises dealing with environmentalism and environmental science, especially pollution, were Arabic medical treatises written by al-Kindi, Qusta ibn Luqa, al-Razi, Ibn Al-Jazzar, al-Tamimi, al-Masihi, Avicenna, Ali ibn Ridwan, Ibn Jumay, Isaac Israeli ben Solomon, Abd-el-latif, Ibn al-Quff, and Ibn al-Nafis.

Their works covered a number of subjects related to pollution such as air pollution, water pollution, soil contamination, municipal solid waste mishandling, and environmental impact assessments of certain localities. Cordoba, al-Andalus also had the first waste containers and waste disposal facilities for litter collection.

ZOOLOGY

In the zoology field of biology, Muslim biologists developed theories on evolution and natural selection which were widely taught in medieval Islamic schools. John William Draper, a contemporary of Charles Darwin, considered the "Mohammedan theory of evolution" to be developed "much farther than we are disposed to do, extending them even to inorganic or mineral things."

According to al-Khazini, ideas on evolution were widespread among "common people" in the Islamic world by the 12th century. The first Muslim biologist to develop a theory on evolution was al-Jahiz (781-869). He wrote on the effects of the environment on the likelihood of an animal to survive, and he first described the struggle for existence and an early form of natural selection. Al-Jahiz was also the first to discuss food chains, and was also an early adherent of environmental determinism, arguing that the environment can determine the physical characteristics of the inhabitants of a certain community and that the origins of different human skin colors is the result of the environment.

Ibn al-Haytham wrote a book in which he argued for evolutionism (although not natural selection), and numerous other Islamic scholars and scientists, such as Ibn Miskawayh, the Brethren of Purity, al-Khazini, Abû Rayhân al-Bîrûnî, Nasir al-Din Tusi, and Ibn Khaldun, discussed and developed these ideas. Translated into Latin, these works began to appear in the West after the Renaissance and appear to have had an impact on Western science.

Ibn Miskawayh's *al-Fawz al-Asghar* and the Brethren of Purity's *Encyclopedia of the Brethren of Purity* (*The Epistles of Ikhwan al-Safa*) expressed evolutionary ideas on how species evolved from matter, into vapor, and then water, then minerals, then plants, then animals, then apes, and then humans. These works were known in Europe and likely had an influence on Darwinism.

CAPITALIST MARKET ECONOMY

Capitalism developed much earlier in Islamic regions than in the Occident. Subhi Y. Labib argues the reason for this was the growing trade economy of the Muslim world, and security from Barbarian invasions. The first market economy and earliest forms of merchant capitalism took root between the 8th–12th centuries in the Caliphate, which are referred to as "Islamic capitalism". A vigorous monetary economy was created on the basis of the expanding levels of circulation of a stable high-value currency (the dinar) and the integration of monetary areas that were previously independent.

Innovative new business techniques and forms of business organisation were introduced by economists, merchants and traders during this time. Such innovations included the earliest trading companies, big businesses, contracts, bills of exchange, long-distance international trade, the first forms of partnership (*mufawada*) such as limited partnerships (*mudaraba*), and the earliest forms of credit, debt, profit, loss, capital (*al-mal*), capital accumulation (*nama al-mal*), circulating capital, capital expenditure, revenue, cheques, promissory notes, trusts and charitable trusts, startup companies, savings accounts, transactional accounts, pawning, loaning, exchange rates, bankers, money changers, ledgers, deposits, assignments, the double-entry bookkeeping system, and lawsuits.

Organizational enterprises similar to corporations independent from the state also existed in the medieval Islamic world, while the agency and aval institutions was also introduced. Many of these early capitalist concepts were adopted and further advanced in medieval Europe from the 13th century onwards.

The systems of contract relied upon by merchants was very effective. Merchants would buy and sell on commission, with money loaned to them by wealthy investors, or a joint investment of several merchants, who were often Muslim, Christian and Jewish. Recently, a collection of documents was found in an Egyptian synagogue shedding a very detailed and human light on the life of medieval Middle Eastern merchants. Business partnerships would be made for many commercial ventures, and bonds of kinship enabled trade networks to form over huge distances.

CROPS

Hundreds of crops were diffused throughout the Islamic world and beyond as a result of the Muslim Agricultural Revolution, some of which include artichokes, bananas, coconut palms, colocasia, cotton, eggplants, hard wheat, lemons, limes, mangos, plantains, rice, sorghum, sour oranges, spinach, sugar cane, and watermelons, among hundreds of other crops.

INDUSTRIAL GROWTH

Muslim engineers in the Islamic world were responsible for numerous innovative industrial uses of hydropower, early industrial uses of tidal power, wind power, and fossil fuels such as petroleum, and the earliest large factory complexes (*tiraz* in Arabic). The industrial uses of watermills in the Islamic world date back to the 7th century, while horizontal-wheeled and vertical-wheeled water mills were both in widespread use since at least the 9th century. A variety of industrial mills were first invented in the Islamic world, including fulling mills, gristmills, hullers, paper mills, sawmills, shipmills, stamp mills, steel mills, sugar mills, tide mills, and windmills. By the 11th century, every province throughout the Islamic world had these industrial mills in operation, from al-Andalus and North Africa to the Middle East and Central Asia. Muslim engineers also invented crankshafts and water turbines, first employed gears in mills and water-raising machines, and pioneered the use of dams as a source of water power, used to provide additional power to watermills and water-raising machines. Such advances made it possible for many industrial tasks that were previously driven by manual labour in ancient times to be mechanized and driven by machinery instead in the medieval Islamic world. The transfer of these technologies to medieval Europe later laid the foundations for the Industrial Revolution in 18th century Europe.

Many industries were generated due to the Muslim Agricultural Revolution, including the earliest industries for agribusiness, astronomical instruments, ceramics, chemicals, distillation technologies, clocks, glass, mechanical hydropowered and wind powered machinery, matting, mosaics, The first large factory complexes (*tiraz*) were built for many of these industries. Knowledge of these industries were later transmitted to medieval Europe, especially during the Latin translations of the 12th century, as well as before and after. For example, the first glass factories in Europe were founded in the 11th century by Egyptian craftsmen in Greece. The agricultural and handicraft industries also experienced high levels of growth during this period.

INDUSTRIAL MILLING

Muslim engineers pioneered two solutions to achieve the maximum output from a water mill. The first solution was to mount them to piers of bridges to take advantage of the increased flow. The second solution was the shipmill, a unique type of water mill powered by water wheels mounted on the sides of ships moored in midstream. This was first employed along the Tigris and Euphrates rivers in 10th century Iraq, where large shipmills made of teak and iron could produce 10 tons of flour from corn every day for the granary in Baghdad. Industrial water mills were also employed in the first large factory complexes built in al-Andalus between the 11th and 13th centuries.

Windmills were first built in Sistan, Afghanistan, from the 7th century. These were vertical axle windmills, which had long vertical driveshafts with rectangle shaped blades. The first windmill was built by the Rashidun caliph Umar (634–44). Made of six to twelve sails covered in reed matting or cloth material, these windmills were used to grind corn and draw up water, and were used in the gristmilling and sugarcane industries.

After paper was introduced into the Islamic world by Chinese prisoners following the Battle of Talas, Muslims made significant improvements to papermaking and built the first paper mills in Baghdad, Iraq, as early as 794. Papermaking was transformed from an art into a major industry as a result. The first fulling mills were later invented in the 10th century, followed by the first stamp mills and steel mills in the 11th century.

The first gristmills were invented by Muslim engineers in the Islamic world, and were used for grinding corn and other seeds to produce meals, and many other industrial uses such as fulling cloth, husking rice, papermaking, pulping sugarcane, and crushing metallic ores before extraction. Gristmills in the Islamic world were often made from both watermills and windmills.

In order to adapt water wheels for gristmilling purposes, cams were used for raising and releasing trip hammers to fall on a material. The first water turbine, which had water wheels with curved blades onto which water flow was directed axially, was also first invented in the Islamic world, and was described in a 9th century Arabic text for use in a watermill.

LABOUR FORCE

The labor force in the caliphate were employed from diverse ethnic and religious backgrounds, while both men and women were involved in diverse occupations and economic activities. Women were employed in a wide range of commercial activities and diverse occupations in the primary sector (as farmers for example), secondary sector (as construction workers, dyers, spinners, etc.) and tertiary sector (as investors, doctors, nurses, presidents of guilds, brokers, peddlers, lenders, scholars, etc.). Muslim women also held a monopoly over certain branches of the textile industry, the largest and most specialized and market-oriented industry at the time, in occupations such as spinning, dying, and embroidery. In comparison, female property rights and wage labour were relatively uncommon in Europe until the Industrial Revolution in the 18th and 19th centuries.

The division of labour was diverse and had been evolving over the centuries. During the 8th–11th centuries, there were on average 63 unique occupations in the primary sector of economic activity (extractive), 697 unique occupations in the secondary sector (manufacturing), and 736 unique occupations in the tertiary sector (service). By the 12th century, the number of unique occupations in the primary sector and secondary sector decreased to 35 and 679 respectively, while the number of unique occupations in the tertiary sector increased to 1,175. These changes in the division of labour reflect the increased mechanization and use of machinery to replace manual labour and the increased standard of living and quality of life of most citizens in the Caliphate.

An economic transition occurred during this period, due to the diversity of the service sector being far greater than any other previous or contemporary society, and the high degree of economic integration between the labour force and the economy. Islamic society also experienced a change in attitude towards manual labour. In previous civilizations such as ancient Greece and in contemporary civilizations such as early medieval Europe, intellectuals saw manual labour in a negative light and looked down on them with contempt. This resulted in technological stagnation as they did not see the need for machinery to replace manual labour. In the Islamic world, however, manual labour was seen in a far more positive light, as intellectuals such as the Brethren of Purity likened them to a participant in the act of creation, while Ibn Khaldun alluded to the benefits of manual labour to the progress of society.

MECHANICAL TECHNOLOGY

The reciprocating suction piston pump with valves and crankshaft-connecting rod mechanism used for raising water, invented by al-Jazari, the father of modern day engineering.

Noria and chain pump (saqiya) machines became more widespread during the Muslim Agricultural Revolution, when Muslim engineers made a number of improvements to the device. These include the first uses of noria and chain pumps for irrigation purposes, and the invention of the flywheel mechanism, used to smooth out the delivery of power from a driving device to a driven machine, which was first invented by Ibn Bassal (fl. 1038–75) of al-Andalus, who pioneered the use of the flywheel in the saqiya and noria.

In 1206, al-Jazari invented a variety of machines for raising water, which were the most efficient in his time, as well as water wheels with cams on their axle used to operate automata. He invented the crankshaft and connecting rod, and employed them in a crank-connecting rod system for two of these water-raising machines. His invention of the crankshaft is considered the most important single mechanical invention after the wheel, as it transforms continuous rotary motion into a linear reciprocating motion, and is central to modern machinery such as the steam engine and the internal combustion engine.

URBANIZATION

There was a significant increase in urbanization during this period, due to numerous scientific advances in fields such as agriculture, hygiene, sanitation, astronomy, medicine and engineering. This also resulted in a rising middle class population. As urbanization increased, Muslim cities grew unregulated, resulting in narrow winding city streets and neighborhoods separated by different ethnic backgrounds and religious affiliations. These qualities proved efficient for transporting goods to and from major commercial centers while preserving the privacy valued by Islamic family life. Suburbs lay just outside the walled city, from wealthy residential communities, to working class semi-slums. City garbage dumps were located far from the city, as were clearly defined cemeteries which were often homes for criminals. A place of prayer was found just near one of the main gates, for religious festivals and public executions. Similarly, Military Training grounds were found near a main gate.

While varying in appearance due to climate and prior local traditions, Islamic cities were almost always dominated by a merchant middle class. Some peoples' loyalty towards their neighborhood was very strong, reflecting ethnicity and religion, while a sense of citizenship was at times uncommon (but not in every case). The extended family provided the foundation for social programmes, business deals, and negotiations with authorities. Part of this economic and social unit were often the tenants of a wealthy landlord.

State power normally focused on Dar al Imara, the governor's office in the citadel. These fortresses towered high above the city built on thousands of years of human settlement. The primary function of the city governor was to provide for defence and to maintain legal order. This system would be responsible for a mixture of autocracy and autonomy within the city. Each neighborhood, and many of the large tenement blocks, elected a representative to deal with urban authorities. These neighborhoods were also expected to organize their young men into a militia providing for protection of their own neighborhoods, and as aid to the professional armies defending the city as a whole.

The head of the family was given the position of authority in his household, although a qadi, or judge was able to negotiate and resolve differences in issues of disagreements within families and between them. The two senior representatives of municipal authority were the qadi and the muhtasib, who held the responsibilities of many issues, including quality of water, maintenance of city streets, containing outbreaks of disease, supervising the markets, and a prompt burial of the dead.

Another aspect of Islamic urban life was waqf, a religious charity directly dealing with the qadi and religious leaders. Through donations, the waqf owned many of the public baths and factories, using the revenue to fund education, and to provide irrigation for orchards outside the city. Following expansion, this system was introduced into Eastern Europe by Ottoman Turks.

While religious foundations of all faiths were tax exempt in the Muslim world, civilians paid their taxes to the urban authorities, soldiers to the superior officer, and landowners to the state treasury. Taxes were also levied on an unmarried man until he was wed. Instead of zakat, the mandatory charity required of Muslims, non-Muslims were required to pay the jizya, a discriminatory religious tax, imposed on Christians and Jews. During the Muslim Conquests of the 7th and 8th centuries conquered populations were given the three choices of either converting to Islam, paying the jizya, or dying by the sword.

Animals brought to the city for slaughter were restricted to areas outside the city, as were any other industries seen as unclean. The more valuable a good was, the closer its market was to the centre of town. Because of this, booksellers and goldsmiths clustered around the main mosque at the heart of the city.

CIVIL ENGINEERING

Many dams, acequia and qanat water supply systems, and "Tribunal of Waters" irrigation systems, were built during the Islamic Golden Age and are still in use today in the Islamic world and in formerly Islamic regions of Europe such as Sicily and the Iberian Peninsula, particularly in the Andalusia, Aragon] and Valencia provinces of Spain. The Arabic systems of irrigation and water distribution were later adopted in the Canary Islands and Americas due to the Spanish and are still used in places like Texas, Mexico, Peru, and Chile.

Muslim cities also had advanced domestic water systems with sewers, public baths, drinking fountains, piped drinking water supplies, and widespread private and public toilet and bathing facilities. By the 10th century,

Cordoba had 700 mosques, 60,000 palaces, and 70 libraries, the largest of which had 600,000 books, while as many as 60,000 treatises, poems, polemics and compilations were published each year in al-Andalus. The library of Cairo had more than 100,000 books, while the library of Tripoli is said to have had as many as three million books. The number of important and original Arabic works on science that have survived is much larger than the combined total of Greek and Latin works on science.

Islamic cities also had an early public health care service. "The extraordinary provision of public bath-houses, complex sanitary systems of drainage (more extensive even than the famous Roman infrastructures), fresh water supplies, and the large and sophisticated urban hospitals, all contributed to the general health of the population."

BRITISH AGRICULTURAL REVOLUTION

The British Agricultural Revolution describes a period of development in Britain between the 17th century and the end of the 19th century, which saw a massive increase in agricultural productivity and net output. This in turn supported unprecedented population growth, freeing up a significant percentage of the workforce, and thereby helped drive the Industrial Revolution. How this came about is not entirely clear. In recent decades, enclosure, mechanization, four-field crop rotation, and selective breeding have been highlighted as primary causes, with credit given to relatively few individuals.

ENCLOSURE

Prior to the 18th century, agriculture had been much the same across Europe since the Middle Ages. The open field system was essentially post-feudal, with each farmer subsistence-cropping strips of land in one of three or four large fields held in common and splitting up the products likewise.

Beginning as early as the 12th century, some of the common fields in Britain were enclosed into individually owned fields, and the process rapidly accelerated in the 15th and 16th centuries. This led to farmers losing their land and their grazing rights, and left many unemployed. In the 16th and 17th centuries, the practice of enclosure was denounced by the Church, and legislation was drawn up against it; but the developments in agricultural mechanization during the 18th century required large, enclosed fields in order to be workable. This led to a series of government acts, culminating finally in the General Enclosure Act of 1801. While farmers received compensation for their strips, it was minimal, and the loss of rights for the rural population led to an increased dependency on the Poor law. Surveying and legal costs weighed heavily on poor farmers, who sometimes even had to sell their share of the land to pay for its being split up. Only a few found work in the (increasingly mechanised) enclosed farms. Most were forced to relocate to the cities to try to find work in the emerging factories of the Industrial Revolution. By the end of the 18th century the process of enclosure was complete.

MECHANIZATION

Jethro Tull made early advancements in agricultural technology with his seed drill (1701) — a mechanical seeder which distributed seeds efficiently across a plot of land. However, he was not the first to invent a seed drill. It took a century after the publication in 1731 of his *Horse hoeing husbandry* for farmers to widely adopt the technology. This may sound to be obvious and necessary processes. But they were not practiced in Europe until the eighteenth century and he was the major contributor in this conversion.

The Chinese were doing this at least by the sixth century BC, and were thus a good 2,200 years in advance of the West in one of the most sensible aspects of agriculture. Contribution of Chinese agricultural methods and tools have been rarely mentioned, rather hidden or kept as minor.

The Chinese iron ploughs, which is generally called 'Dutch design' because the Dutch brought them from China, contributed drastic improvement in productivity not only in Britain but also in Europe and the U.S. Tull's seed drill was inspired by the excellent Chinese counterpart, which never arrived in Europe. The Chinese rotary winnowing fan revolutionized winnowing process since its introduction to Europe.

Joseph Foljambe's Rotherham plough of 1730, while not the first iron plough, was the first iron plough to have any commercial success, combining an earlier Dutch design with a number of technological innovations. Its fittings and coulter were made of iron and the mouldboard and share were covered with an iron plate making it lighter to pull and more controllable than previous ploughs.

It remained in use in Britain until the development of the tractor. It was followed by John Small of Doncaster and Berwickshire in 1763, whose 'Scots Plough' used an improved cast iron shape to turn the soil more effectively with less draft, wear, or strain on the ploughing team.

Andrew Meikle's threshing machine of 1786 was the final straw for many farm labourers, and led to the 1830 agricultural rebellion of Captain Swing (a probably mythical character comparable to the Luddite's Ned Ludd).

In the 1850s and '60s John Fowler, an agricultural engineer, pioneered the use of steam engines for ploughing and digging drainage channels.

FOUR-FIELD CROP ROTATION

During the Middle Ages, the open field system had employed a three year crop rotation, with a different crop in each of the three fields, eg. wheat and barley in two, with the third fallow. 'Fallow' is a term which means that the field is empty, there is nothing growing there. Over the following two centuries, the regular planting of nitrogen-rich legumes in the fields which were previously fallow slowly increased the fertility of croplands. The planting of legumes (leguminosae, plants of the pea/bean family) helped to increase plant growth in the empty field because they used a different set of nutrients to grow than the grains. The legumes put back nutrients the grains used, nitrates produced from nitrogen in the atmosphere, and the grains put back the minerals the legumes used. In a way, they fed each other. Other crops that were occasionally grown were flax, and members of the mustard family. Medieval record keepers did not distinguish between rape seed or other mustards grown for animal feed and mustard grown for mustard seed for condiments. When the pastures were brought back into crop production after their long fallow, their fertility was much greater than they had been in medieval times. The farmers in Flanders (current day Belgium), however, discovered a still more effective four-field rotation system, introducing turnips and clover to replace the fallow year. Clover was both an ideal fodder crop, and it actually improved grain yields in the following year (clover is part of the pea family, leguminosae). The improved grain production simultaneously increased livestock production. Farmers could grow more livestock because there was more food, and manure was an excellent fertilizer, so they could have even more productive crops. Charles Townshend learned the four field system from Flanders and introduced it to Great Britain in 1730.

SELECTIVE BREEDING

In England, Robert Bakewell and Thomas Coke introduced selective breeding (mating together two animals with particularly desirable characteristics), and inbreeding (to stabilize certain qualities) in order to reduce genetic diversity in desirable animals programmes from the mid 18th century. Robert Bakewell cross-bred the Lincoln and Longhorn sheep to produce the New Leicester variety. These methods proved successful in the production of larger and more profitable livestock.

TECHNOLOGY FROM FLANDERS

The British Agricultural Revolution was sparked in part by advancements in Flanders, including the aforementioned four-crop rotation. Due to the large and dense population of Flanders, which forced farmers to take advantage of every inch of usable land, the country had become a pioneer in drainage and reclamation technology. Many Flemish experts went to The Netherlands and reclaimed land there. Finally, Dutch experts like Cornelius Vermuyden brought the technology to Britain.

EFFECTS ON HISTORY

The Agricultural Revolution in Britain proved to be a major turning point in history. The population in 1750 reached the level of 5.7 million. This had happened before: in around 1300 and again in 1650. Each time, the appropriate agricultural infrastructure to support a population this high was not present, and the population fell. However, by 1750, when the population reached this level again, an onset in agricultural technology and new methodology allowed the population growth to be sustained.

The increase in population led to more demand from the people for goods such as clothing. A new class of landless labourers, products of enclosure, provided the basis for cottage industry, a stepping stone to the Industrial Revolution. To supply continually growing demand, shrewd businessmen began to pioneer new technology to meet demand from the people. This led to the first industrial factories. People who once were farmers moved to large cities to get jobs in the factories. It should be noted that the British Agricultural Revolution not only made the population increase possible, but also increased the yield per agricultural worker, meaning that a larger percentage of the population could work in these new, post-Agricultural Revolution jobs.

The British Agricultural Revolution was the cause of drastic changes in the lives of British women. Before the Agricultural Revolution, women worked alongside their husbands in the fields and were an active part of farming. The increased efficiency of the new machinery, along with the fact that this new machinery was often heavier and difficult for a woman to wield, made this unnecessary and impractical, and women were relegated to other roles in society. To supplement the family's income, many went into cottage industries. Others became domestic servants or were forced into professions such as prostitution. The new, limited roles of women, dubbed by one historian as "this defamation of women workers", (Valenze) fueled prejudices of women only being fit to work in the home, and also effectively separated them from the new, mechanized areas of work, leading to a divide in the pay between men and women.

Towards the end of the 19th century, the substantial gains in British agricultural productivity were rapidly offset by competition from cheaper imports, made possible by advances in transportation, refrigeration, and many other technologies. From that point, farming in Britain entered a period of economic struggle which continues to the present day.

URBAN FARMING

HISTORY OF URBAN FARMING

Urban farming is not a new idea in the United States. During World Wars I and II, the government urged those not living in rural areas to plant food in their yards. This relieved pressure on traditional agriculture, and allowed for the production of enough food to support the war efforts. These so called, "Victory gardens," were very successful, growing roughly 40% of the vegetables consumed in the U.S. in 1942 and 1943.

Urban Farms Today

Urban farming today is being done on all scales. It can vary from large farms of several acres in size, to the smallest which can consist of a small raised bed. New and bold urban farming initiatives are starting to begin. Roof top gardens are beginning to replace the hot black roofs of businesses in major cities. Vertical farms are seen as the future, producing the food where the demand is greatest.

Urban farming is happening all across the world in major cities. Recently, the urban farming movement has a taken off in American cities especially those hit hardest with economic downfalls. The farm depicted is using earth boxes to circumvent the common problem of contaminated soil in urban areas. Newark with a total of 1,900 gardens on 30 acres produced approximately $915,000 of food value in one year.

Benefits

Truth is most people don't see farming as an opportunity to make money. Even fewer people believe that they have the skills to farm themselves. Both of these assumptions are false when it comes to urban farming. For a homeowner to start a garden in their backyard it takes a shovel and some hard work. Aside from the seed and fertilizer these types of small backyard personal gardens require little to no money. If the correct crops are chosen the profits can be substantial. An example of a successful urban farm project comes from the outskirts of Philadelphia in 2006. A sub-acre garden hauled in $67,000 from crops like salad greens and baby vegetables. This type of income can revolutionize the economic outlook of poorer communities. Urban farming doesn't have to be a hobby on the side to make extra cash; it can actually become a full time job with excellent pay. Kenny Penzien, a second generation farmer has successfully harvested and sold his produce for thirty years. Although Mr. Penzien is not an urban farmer, he currently cultivates 80 acres of land; he was very helpful in answering questions about the feasibility and economics of urban farming. Mr. Penzien stated the following during my interview with him.

"Urban farming has a tremendous advantage over shipped foods, especially when your customers are located in that area. I currently sell my produce locally and at the Eastern Market in Detroit. It isn't uncommon for me to make a trip everyday to deliver produce to the market. If I could save 140 miles worth of gas a day I would. So in this respect the urban farming has the advantage."

The fact that urban farms are close to the intended customers gives them a superior advantage over the traditional produce found at grocery stores. Urban farms will be able to provide a cheaper product due to savings in transport costs and a fresher more desirable product. Meaning the urban farmers will be able to make more profits.

The profitability of a farm is tightly correlated with the types of crops grown and proper growing methods. When space is limited, it must used to its full potential. Mr. Penzien stressed the importance of multiple crops, getting three or four growing cycles in a growing season. Crops that achieve the best results are herbs, many types of lettuce, summer squash, peas, flowers, and fruit. This is unlike the traditional, one crop per season mentality of larger scale rural farmers. Multiple crops increases revenue and lessen the financial impact if one crop was to fail. Mr. Penzien has a green house where he starts plants before the snow is even melted. He recommended that if a green house is impractical starting a couple flats of plants inside your porch will give urban farmers a head start. The goal is get a produce before it is in season. With a high demand and low supply, successful farmers will be able to charge a higher price for their produce.

Spin Farming is an organization that specializes in small plot intensive gardening. They have shown that if one acre of land is intensively relay cropped (three or more harvests a year) with high value crops that it can yield 120,000 dollars a year. If done properly through the use of correct crop selection and proper farming methods urban farming can become an economic force in any city.

Healthier Food, Healthier Populace

Limited income households tend to purchase bulk foods that fill them up, while neglecting the fruits and vegetables that are full of nutrients. It has been shown that a 5-10 day transportation and storage lag between production and consumption leads to losses of 30-50% in nutrients. So not only do lower income families eat minimal fruits and vegetables, the produce available to them is lacking in nutrients due to a lack of freshness.

The deficiency of a healthy diet as made obesity, diabetes, and heart disease rampant among the lower income households located within cities. Urban farming is an obvious answer to solving unhealthy diets. A garden the size of 10x10 meter plot can provide most of a household's total yearly vegetable needs, including much of the household's nutritional requirements for vitamins A, C, and B complex and iron. Studies have shown that an urban garden will yield six dollars for every one dollar invested. Urban gardens free up money for non-food items and can easily provide for a healthy diet. Even if a family doesn't choose to grow their own garden, the hope is that their neighbors will have gardens. Allowing trading or purchasing of produce they need to obtain a healthy diet. A healthy side-effect of this scenario is community building. With neighbors relying on each other and doing business, citizens will take pride and build a healthier self-image of their community.

Urban farms will help supply food banks of much needed fresh produce which often is not donated. Church groups and schools are starting their own urban farms to supply food for these organizations. Let's not forget that farming is hard work. Anyone that participates in farming will get a good workout, especially if you choose the cheaper non-mechanical method of tending the garden. Fresh produce grown locally is benefiting the health and helping to grow a closer knit community.

City Beatification

Cities hit hard by the collapse of manufacturing are littered with decaying factories, abandoned homes, and unutilized green spaces. These spaces only aid in the degradation of a citizens image of the city and speed up the population losses that hinder many cities. Urban farming is flexible and can turn almost any space into a garden. From an abandoned factory roof, an empty lot, to even a paved parking lot can be converted into a useful space that benefits the community. Ron Doetch an executive director of the Michael Fields Agricultural Institute realizes the problem with vacant lots and states, "Unless all this land is going to be redeveloped by local housing authorities, it is very likely that most of the vacant land in the most devastated central city neighborhoods will remain vacant 10 years from now". Cities need to be more realistic about the long-term prospect that vacant lots in communities are not going to be redeveloped. Mr. Doetch also indicates that green space and gardens can significantly improve housing values. So the solution to vacant land is really quite simple. Where would you rather invest: in a neighborhood filled with trash filled vacant lots or where residents are out tending the neighborhood garden? I chose the latter option.

Reduced Ecological Footprint

The cost of producing and transporting food from the far reaches of the world is very costly in terms of fossil fuels and pollution. As populations increase and continue to shift towards metropolitan regions the rate and distances food will travel will increase. Today the average distance food travels from field to grocery store is 1500 miles. Another concern is that traditional rural agriculture will not be able to keep pace will global food demand. This could result in a global food crisis, forcing the U.S to ship food even farther to meet city food demands.

Urban farming provides a solution where transportation of produce is virtually eliminated and all the while providing produce on reclaimed land. In 2003, the United States consumed approximately 20 million barrels of oil per day of which two-thirds was for transport (Brown, 2008). If a fraction of that demand could be cut, it would be a great step forward in lowering pollution and controlling energy needs in the future. Also, as cities expand into fertile and productive farmland that land will become virtually unproductive. Urban farms use of underutilized land in the heart of the metro regions will offset the loss of productive farm land.

Challenges Associated with Urban Agriculture

Urban farming does have its challenges hindering it fullest potentials. The first challenge facing urban farming is the economic state of the cities, the availability of land and cost, site controls, environmental contamination, theft and vandalism, and the availability of credit and investment capital.

Some cities are just not friendly to urban farms. Zoning policies in certain cities will restrict the idea of farms using public land for these purposes. This combined with a poor economy where investment capital is slim will further hinder the purchase of any land for an urban farm. I feel that the only solution to these two problems is the small backyard gardens. Where zoning and public lands are not needed and the scale of the farm is not large enough to need financing. Many urban farmers run into the problem of soil contamination. It is crucial to test soils for lead and other heavy metals that can be absorbed by produce and then consumed. This problem seems to occur in manufacturing cities where pollution has occurred for extended periods of time. The use of earth boxes like those seen in Figure 1 can bypass this problem. All though the use of earth boxes increases the initial investment of an urban farm it doesn't hinder its productivity.

There are other potential risks with urban gardening and farming that will require common sense strategies. Standing water can attract bugs, including mosquitoes that can carry diseases. So proper care of water catch systems need to be properly covered or a use of public must be used to limit this problem. Improper use of animal manure for fertilizer can spread diseases but this is averted by proper composting methods before placement on gardens.

Detroit has Land and a Friendly Local Government

Empty lots and burnt-out houses have surpassed the number of livable homes in some neighborhoods. Detroit has 70,000 vacant lots, comprising 27% of its 139 square miles. That is over 27 square miles of vacant land currently residing in Detroit unused. These large expanses of land will allow for a much larger sized farm that would be unattainable in other cities. This potentially could allow Detroit to become more successful than any other urban farming project. Not only does Detroit have vast amounts of free land but the city actually wants to have farming. In 2008, a land bank of 10,000 city owned properties was created to oversee the leasing of properties. The properties can be leased through the Garden for Growth programme for a small fee of 50 dollars. The city government's acceptance of urban farming opens the doors for any prospective farmer to obtain a piece of land easily.

Detroit's Problems suit Urban Farming

Michigan has one of the highest unemployment rates in all of the United States. Detroit being an auto industry centered city has had drastic amounts of layoffs in the recent years. The city has suffered over a million resident and hundreds of thousands of jobs since the early 1960s. Urban farming could start a new economy in Detroit and help diversify its economy.

Detroit's poverty and unemployment rates are a toll on Detroit's citizens. Health problems ranging from obesity, hypertension, diabetes, and heart disease are common in poorer communities. This is a direct result of fast food and processed foods. Kami Pothukuchi associate professor of geography and urban planning states, "That Detroit has only a few grocery stores where residents can buy a variety of nutritious, high quality foods".

Therefore healthy food is located outside the city and is unobtainable to lower class residences. Urban farming would create a pathway to improve diets and correct health problems.

Also, it will reduce the ecological impact of Detroiters having to travel many miles outside of the city to get fresh quality produce. Detroit is 27% empty. That empty land is fragmenting neighborhoods. Urban farming will bridge the gaps between homes and help patch together communities that have nearly disappeared. The gardens will help bring property values up and encourage investment back to the city.

Detroit's Eastern Market is the perfect venue to sell produce and establish a name for urban farming. It consists of 43 acres and attracts on average 45,000 people on Saturdays (Flisram, 2009). This is essential if urban farms are to make profits and to noticeably impact the economics of the city. Also, the Eastern Market is a perfect meeting spot and education centre to attract more potential farmers to insure an enduring urban farming movement.

PARTICIPATORY APPROACHES IN ANALYSIS OF URBAN AGRICULTURE

Participatory approaches are well established within the tradition of participatory rural appraisal (PRA) and participatory research. There are fewer examples of use and critical assessment of participatory methods in the urban agricultural context, despite some important advantages they offer (Garrett, 2001). Engagement in a participatory process encourages participation and empowerment of local communities and prepares the ground for improved local governance. It facilitates the integration of gender, cultural and environment aspects in the development of project proposals, helping to avoid a misfit between intervention and need. It also helps to capture complexity and permit better problem definition.

Santandreu presents interesting experiences using a "Rapid Visual Appraisal" (RVA) approach to go beyond the descriptive component of many case studies (Santandreu, (2001). The RVA has its theoretical and methodological background in Social ecology, Participatory Rural Appraisal and Rapid Rural Evaluation as well as in Landscape ecology. In Santandreu's opinion, the application of an RVA leads to the collection and synthesis of important information regarding UA. It also facilitates the engagement of multi-sectoral stakeholders in a process which can lead into further action. A key point of the RVA is that the situation analysis (appraisal phase) leads into action, dissemination and enhanced participation. The different phases of the RVA are as follows:

- Conceptualisation of the topic
- Conceptualisation of the context
- Conceptualisation of information
- Visual appraisal
- Participatory interviews
- Formulation of participatory action plan.

Santandreu concludes that the integration of different stakeholder groups in the process of analysing and evaluating the local UA context contributes to the valorisation of local practices as well as to the development of new knowledge. Additionally, the engagement in a participatory process prepares the ground for improved local governance and the integration of gender, cultural and environment aspects in the development of project proposals.

Gabel (2001) describes her fieldwork experiences in Harare, Zimbabwe, using participatory and action-oriented research approaches to investigate and address the food provision needs and strategies of low income households. Her particular focus was on the role of women in food provisioning and their use of urban cultivation, taking into account local level policy influences. Her choice of participatory and gender based methodologies reflect the priority given to critical reflection on the role of the researcher and relationships with participants.

The use of PRA tools such as transect walks and mapping of plots and land use are usually recommended as introductory participatory exercises which provide researchers with a initial overview of the area and participants with the opportunity to familiarise themselves with the research and to take a lead in identifying which issues are important to them. However, in high density urban areas with an influx of new residents and where access to land is limited and often highly politicised, this might arouse mistrust (Gabel, 2001). Therefore, it is important to build up trust first with several key informants before issues such as land use and tenure can be explored in depth.

Time is needed for building trust within the participatory analysis process. Urban locations often have the advantage of accessibility, but access to individuals is complex because of the diversity of livelihoods – people engage in multiple activities in different locations and many combinations exist among household members. Norton (1994, p55), highlights some of the problems of applying PRA in an urban context, as the ideas behind PRA are often based on assumptions relevant to the rural context. For example:

- *the assumption of mutual knowledge*: This is relevant to methods which rely on detailed knowledge of the situation of other community members, in particular wealth ranking. It is often difficult for urban key informants to rank their community due to insufficient knowledge. A comment from a respondent in Ghana illustrates this: "only if neighbours are in serious difficulties can one see the social resources in terms of networks of kin and friends they can draw upon". In rural areas these aspects of networks are usually more visible and publicly known.
- *the assumption of homogeneity in livelihood patterns*: This applies especially to seasonal diagramming. Urban livelihoods and incomes are also subject to seasonality, but patterns and effects are likely to be very varied. There is a need to rethink the likely lines and patterns of seasonal change and how these will affect different social and livelihood groups. The timing of studies needs to accommodate seasonality, day/ evening/ weekday work patterns of different stakeholder groups.
- *The assumption of community*: the understanding of what constitutes a community is more variable because of the diversity of the urban population. The complexity of social identification and affiliation is likely to be greater in the urban context, compared with the rural.

Participatory research methods can be used in an urban context, but it is important to maintain critical awareness of these potentially misleading assumptions. Mitlin and Thompson (1994), also discuss differences between conducting PRA in urban and rural contexts associated with the greater diversity in urban livelihood strategies, the extent of reliance on natural resources, the combination of formal and informal activities, in language and ethnicity, and in tenure arrangements for housing and land, (*e.g.* renting, illegal occupation etc.). Local government, although often weak, plays a role in intervention in urban areas, for example with respect to squatters and zoning of land. However, despite the differences between urban and rural contexts, Mitlin and Thompson argue that they still have much in common and the linkages between urban and rural should be considered. However, they underscore the importance of developing a better understanding of the added contribution that participatory tools and methods might offer in urban areas.

The term 'participation' can mean many different things to different people. Approaches which seek to go beyond mere consultation face the question of who is able to participate and how to ensure equity of participation. This links with the need for stakeholder analysis discussed below. Tools are needed to encourage the direct participation of urban agricultural producers, and for consultation with secondary stakeholders such as local leaders, politicians, professional workers in health, education, agriculture, traders and major employers.

The obstacles to participation, especially in the poorest areas, include low levels of social capital, poverty and limited access to assets, pressure of work, insecurity resulting from crime or political activity and social isolation. There may be constraints on the participation of the poor because of the power of dominant groups. Beacon Mbiba points out that "critics have been quick to observe that the "participatory" ideas in communicative-collaborative planning have created more practical problems for the poor; more meetings leading to burn-out and disillusionment among participants.... In reality, the ability to participate is mediated by assets at the disposal of each potential participant.'.

There may be trade-offs between encouraging participation and achieving a goal. "Weighing up the pros and cons between empowerment of those living in poverty and pragmatic involvement of influential stakeholders is an important consideration which will determine the success or otherwise of the strategy." Local government structures may or may not provide a supportive infrastructure.

Whether participatory methods are used or not, depends on the objectives of the study, the type of analysis and for whom the analysis is done. There is a distinction between the processes of setting objectives and accounting for results, and the actual methods of enquiry. Either, neither, or both, may be participatory.

The decision making process can be democratic and accountable whether or not the methods of enquiry are themselves participatory. A purpose may be arrived at with participation but the kinds of interaction may differ, particularly if addressing issues of public policy. Some scientific methods have to be participatory – *e.g.* studies of attitudes, studies of social and political processes.

Participatory methods are useful to explore a diversity of perspectives and definitions. Their use can be a pragmatic choice to improve data collection and findings or may be motivate by a desire to promote empowerment. A combination of participatory and non participatory methods is possible *e.g.* entomological monitoring by experts to address a need for solutions to pest problems defined through participatory diagnosis; or participatory nutritional record keeping by households as part of a government inspired nutritional survey. Participatory methods do not necessarily imply qualitative data, nor are non participatory methods necessarily quantitative.

ANALYSIS OF DYNAMICS AND POVERTY

Rapid change is characteristic of many urban environments, hence there is a need for practical methodologies to capture trends and dynamics influencing urban and peri urban agriculture, at different scale levels (intra household, group/neighbourhood, city). Urban agriculture is often studied from a rather narrow static perspective (descriptions of urban farming systems, benefits and constraints, economic analysis of its contribution etc.). Often it is not linked to wider urban livelihoods issues and the underlying dynamics in terms of changing land tenure and emerging land markets, rural-urban linkages, migration patterns, local perceptions, the national economy, urban development policies, environmental issues and the historical, institutional and political context. Mbiba (2001a) highlights the importance of understanding the context and diversity of local dynamics, especially concerning land, governance and institutional action to resolve associated conflicts.

An important characteristic of poverty is its dynamism; poor individuals and communities are not necessarily permanently poor and it is important to distinguish between chronic and transient poverty (Rakodi, 1998). People may move in and out of poverty as a result of wider shocks and stresses, seasonality (in urban areas, health, food prices and vulnerability to flooding, or time of school fees), cyclical factors, *i.e.* monthly wage payments, life cycle factors, both intra and inter-generational.

To understand the wider dynamic and institutional context, it is important to explore what changes people have perceived in relation to their livelihoods and to urban agriculture, how they have responded and their perceptions and aspirations. Visualisation using photographs has been used to explore values and perceptions relating to urban agriculture and urban development in general.

Photographs were used in individual interviews to explore local perceptions of urban development. Respondents were presented with different sets of photographs in groups of three and had to explain the differences/similarities.

These triadic comparisons and ranking procedures revealed cognitive patterns of thinking about urban areas. The advantage is that many people are attracted to photographs and like to talk about what they see. Photographs show a great deal of detail that can be compared, but may also be interpreted differently by respondents. The use of photographs also has the advantage that general value orientations based on observable traits can be elicited rather than an evaluation of a particular location.

ASSESSMENT OF FOOD SECURITY AND NUTRITIONAL STATUS

Urban agriculture makes a direct and indirect contribution to urban food supply. Data on indirect contribution to urban food supply can be collected through household surveys asking the respondents to estimate the proportion of the home consumption that is covered by home cultivation (Moustier, 2001). A comparison of weekly expenditure on vegetables, meat, fish and eggs among farming and non-farming households also indicates indirectly the value of urban agriculture at the household level (Martin et al, 2000). Moustier suggests that the direct contribution to urban food supply is estimated by the quantities of different produce traded at the market, or alternatively by measuring the total harvest (average areas x average yield) which is complicated by the practice of intercropping and piece meal harvesting as well as seasonality. A way around that where maps are available, is to randomly select plots, covering different crops, estimate the harvest through close monitoring with the farmers and extrapolate these data.

Studies measuring impact of urban agriculture on food security tend to support the hypothesis that urban agriculture improves the food security of vulnerable households.

However, relatively few studies attempt to measure the link with nutrition. FAO have been developing guidelines for the implementation of national food insecurity and vulnerability information and mapping systems (FIVIMS).

It includes a conceptual framework for understanding possible causes of low food consumption and poor nutritional status, *e.g.* socio-economic and political environment, performance of the food economy, care practices and health and sanitation. The need has been identified for locally collected information on the situation of food insecure and vulnerable individuals and households.

Pederson (2001) examines the applicability of participatory approaches to planning intervention programmes for improving urban food and nutrition security. He comments that policies and programmes have usually been designed after the definition of nutritional problems by experts. The underlying socio-economic and cultural factors are often ignored, although these interact with biological processes creating nutritional and health variability within subgroups.

In contrast, bottom-up involvement helps to accelerate nutritional improvements and other benefits. There is complementarity between formal nutrition surveys in which overall nutritional status is accurately measured, and participatory methods for understanding interactions and behaviour. Participatory approaches are effective in revealing the complexity of issues affecting nutrition at community level.

Pederson describes a community mapping tool kit which is designed to enable local communities to understand their local food economy and provide information to lobby government and improve food and nutrition security. It involves a multidisciplinary, critical examination of the causes and consequences and alternatives for improved food security in local context leading to action planning, involving major stakeholders. It is designed in 3 phases with follow-up activities; a pre-planning phase includes stakeholder analysis of those directly affected by food and nutrition policy and who would benefit from interventions to improve food security, as well as indirect stakeholders whose cooperation is essential to ensure strategies to improve nutrition are successful; a preliminary planning phase of group data collection for a socio-economic profile and a community map to visualise the factors and conditions affecting food security and their interaction. Problems and assets were identified and problems were ranked by prevalence and severity. The third phase of action planning is followed up by periodic reappraisal, information sharing within and beyond community, food action projects, and policy influence.

Participatory approaches supplement conventional surveys measuring food intake and nutritional status by offering an understanding into the complexities of food and nutrition insecurity in the community.

6

Poverty Reduction and Food Security

BIOTECHNOLOGY AND POVERTY REDUCTION AND FOOD SECURITY

Agricultural biotechnology will contribute to poverty reduction and food security if scientists can develop technologies to increase quality and yields of food crops, and the technologies are adopted by small farmers. For this to happen, biotechnology R&D will have to meet four conditions:

- It must address both the problems of small farmers in rainfed areas where most of the poor live, and those of small farmers in irrigated areas, which provide the bulk of food grain supply in Asia.
- It must focus on crops, livestock, and fish commonly grown by small farmers. Major crops are rice, tropical maize, wheat, sorghum, millet, banana, cassava, groundnut, oilseed, potato, and sweetpotato. Biotechnology R&D should also focus on high value cash crops (*e.g.*, cotton, soybean, and vegetables) that can increase the incomes of small farmers through crop diversification. The prospect for improving these crops is bright due to the large demand for them in urban areas and in international markets. Fish and livestock (cattle, sheep, goats, pigs, and chickens) are also important.
- The technology to be developed and delivered to small farmers must be simple, low cost, and carry little or no risks to human health and the environment. As in the case of the Green Revolution, the most effective strategy to increase food production is through improved seeds that possess high yield potential, fertilizer responsiveness, resistance to pests and diseases, good agronomic characteristics, and good nutritional quality.
- Biotechnology development should be accompanied by favorable policy environment; good governance; investments in rural

infrastructure; agricultural research and extension; and credit and marketing.

INVEST IN AGRICULTURAL BIOTECHNOLOGY

In much of Asia. yields of major food grains are stagnant or declining in the face of population increases. Pests and diseases cause substantial preharvest and postharvest losses of crops, livestock, and fish. Solutions to many of these problems may lie in the various applications of modern biotechnology. The use of molecular markers to tag specific traits is accelerating the breeding of new varieties of plants and animals. New understanding of plant and animal genes through genomics may offer ways of increasing crop yields.

These new developments when used in conjunction with developments in the physical and social sciences, offer more sustainable means for obtaining necessary productivity increases that are less dependent on environmentally damaging inputs of chemical fertilizers and pesticides. Given appropriate policies and necessary human and financial resources, modern biotechnology could make an extremely important contribution to future agricultural growth.

During the next 25 years, Asia will need a Second Green Revolution, often called Biorevolution or Doubly Green Revolution. Conway (1997) pointed out that the next technology-driven revolution must be doubly green—it must increase food production at a faster rate than in recent years without significantly damaging the environment. It must also increase incomes and increase access to food by the poor. Compared to the Green Revolution of the 1970s, Biorevolution will be characterized by the following features:

- Potentially many crops (particularly high value and specialty crops), will be affected as well as livestock and aquaculture.
- Potentially all areas, both irrigated and rainfed, will benefit from biotechnology R&D.
- Technology development and dissemination will substantially involve the private sector with the public sector playing the role of facilitator and regulator.
- Many processes and products will be patentable and protectable.
- Capital costs of research will be high.
- Molecular and cell biology expertise will be required in addition to expertise in conventional plant breeding and other agricultural sciences.
- Modern biotechnology (genetic engineering) is not a silver bullet for achieving food security, but used in conjunction with other techniques it may be a powerful tool in the fight against poverty and food insecurity. Other approaches are available and should be used. Narrowing the yield gap between those obtained from farmers' fields and those from experiment stations using the current

technologies is just one example. However, there is concern that some conventional alternatives will not be able to produce the desired results within a limited time. The advantage of modern biotechnology rests on the speed at which desired crop varieties are produced. In some cases, the desirable genetic combination of traits is simply not possible through common breeding methods, and can be done only through genetic engineering. To increase food production by at least 40 percent within the next 25 years, Asian countries not only have to move towards the best technological frontier (to push farmers' yields to the optimum level), but keep moving the technological frontier itself. As long as product safety, environmental and ethical concerns, and IP issues are adequately addressed, modern agricultural biotechnology has the potential to significantly increase the quantity and quality of the food supply for developing countries.

APPLYING BIOTECHNOLOGY IN RESOURCE POOR AREAS

The resource-poor, rainfed areas in Asia are home to many poor people, and their population is growing rapidly. Although migration may sometimes be the only viable livelihood strategy, in many such areas sustainable intensification of agriculture may be the best way to achieve poverty reduction and food security. Development policy has often neglected these areas characterized by poor soils, shorter growing seasons, lower and uncertain rainfall, and little infrastructure or access to markets. Yield increases usually lag behind population growth. Efforts by poor farmers to expand cultivation onto new lands to eke out survival often lead to deforestation and erosion of fragile highland soils. That in turn threatens hydropower, road, and irrigation infrastructure in the lowlands. Policies and investments are required that can achieve food security in ways that protect natural resources, thereby breaking the vicious cycle of poverty, low productivity, and environmental degradation. Creation of nonfarm, rural jobs will also be important (ADB 2000b).

Sizeable investments in biotechnology in Asia have been made in rainfed areas, although DMCs continue to give high priority to irrigated agriculture. Irrigated areas produce 75 percent of total cereal production in Asia. Many IARCs such as the Asian Vegetable Research and Development Centre, CIMMYT, CIP, ICRISAT, ILRI, and IRRI; and most DMCs have invested heavily in R&D on orphan crops such as rice, tropical maize, sorghum, potato, banana, groundnut, and sweetpotato.

APPLICABILITY OF BIOTECHNOLOGY TO SPECIFIC AGRICULTURAL OBJECTIVES

In Asia, biotechnology has been applied mainly to develop improved varieties adapted to specific environments. Specific areas where new applications of biotechnology could address poverty reduction and food security are summarized as follows:

- Increasing Productivity and Stability of Crops in Rainfed and Marginal Environments. Producing more food on the same area of cultivated land would reduce pressure to expand cultivated areas to forests and marginal areas. Broadening tolerance of existing HYV cereals for drought, flooding, salinity, heavy metals, and other abiotic and biotic stresses would increase yields in rainfed areas. CIMMYT and IRRI, in cooperation with NARSs in different countries, are focusing their efforts to develop HYVs of rice and maize for rainfed areas.
- Improving Water Use Efficiency in Crops. Future availability of water for agriculture is a major issue. By 2020, there may be a water crisis in Asian agriculture. Crops with higher water use efficiency and tolerance for drought would be an advantage. With the completion of the rice genome, scientists will be able to identify genes responsible for drought tolerance.
- Integrated Pest Management (IPM) Strategies to Reduce Pesticide Use. The insecticides now used in many developing countries are often older, broad spectrum, acutely toxic compounds. Many are now banned in industrial countries except for export. They are a significant health hazard to farmers and farm workers throughout Asia. Over 80 percent of total pesticide use in Asia is on cotton, rice, and vegetables. In the past, governments encouraged the use of chemical pesticides as part of a yield-increasing package of farm inputs, and some governments still subsidize pesticides. The need is now recognized for more sustainable approaches to IPM to reduce losses to pests without harmful side effects on human health and the environment. New pest-and diseaseresistant crops, produced with the aid of modern biotechnology, may be important components of IPM strategies.
- Early applications are being seen in the use of novel sources of resistance (Bt genes) to control insect pests on maize and cotton. New pest-resistant cotton varieties are being grown widely in the PRC by 3 million farmers covering 500,000 ha. Similar cotton varieties are undergoing field tests in India, Indonesia, and Thailand. Genetically modified maize resistant to insect pests is being tested in the Philippines.
- Increasing Disease Resistance in Crops. Biotechnology applied to disease resistance in crops would be especially valuable for combating diseases in tropical environments that have not been controlled by conventional means. These include important diseases such as downy mildew of maize; bacterial blight and blast of rice,

papaya ringspot virus; banana bunchy top virus; and bacterial wilt of tomato, eggplant, and potato.

- Increasing Nutritional Quality. Enhancing the protein, vitamin, and micronutrient contents of food grains would greatly benefit poor consumers who cannot afford supplementary vitamins and micronutrients. Under the Golden Rice Project, IRRI scientists are undertaking biotechnology research to enhance the vitamin A content of rice by introducing genes from daffodil.
- Increasing Sustainable Production of Livestock. Research to increase the productivity and quality of farm animals, especially of dairy cows, small ruminants (sheep and goats), and water buffalo is promising. Vaccines and disease diagnostics for control of epidemic and endemic diseases of livestock are being developed.
- Increasing Productivity in Fisheries and Aquaculture. Development of vaccines and disease diagnostics for control of epidemic and endemic diseases of shrimp in Thailand is a good example of where modern biotechnology has been used successfully.

BENEFITS FROM AGRICULTURAL BIOTECHNOLOGY

Realized or potential benefits of agricultural biotechnology can be categorized as economic, social, and environmental. A large share of the benefits are concentrated in industrialized countries, where a diversity of applications are widespread in human and animal health care and in many aspects of food production and processing.

As the first transgenic biotechnology product was not commercialized until 1996, little quantitative analysis has been publicly available until recently. The studies published by ISAAA have helped fill the gap, at least with respect to transgenic crops. ISAAA analyzed the global status of transgenic crops, including the area planted and the value of the market for transgenic seed, which was estimated between $2.7 billion and $3.0 billion in 1999.

The studies show that the economic benefits were fairly evenly distributed between the companies that developed the technology and the farmers growing transgenic crops. ISAAA studies have also demonstrated positive environmental impacts. In the cases of herbicide-tolerant soybean and herbicide-tolerant canola, herbicide use was considerably reduced even though weeds were better controlled, and yields increased significantly. In the cases of Bt maize, cotton, and potatoes, targeted insect pests were effectively controlled and yields were increased. Insecticide use for nontargeted insects was also reduced.

Economic Benefits in Developing Countries

Few published analytical studies have attempted to assess the impact of biotechnology in developing countries. Ex-ante studies on transgenic pest-or virus-resistant sweetpotato in Kenya and on transgenic virus-resistant potato in Mexico suggest substantial benefits for both producers and consumers. They point out, however, that smallholders will fully realize benefits only with improved farm-level management and more efficient seed distribution.

The introduction of pathogen-free banana planting material in Kenya illustrates the benefits that accrue to small-scale farmers, mainly woman. Yield increases from the new material have been substantial; in some cases yields almost doubled. This example is significant in that it involved community groups and farmers in the development and field-testing. By the time the new planting material was available commercially, the benefits of the technology had already been demonstrated to farmers, with the result that initially demand exceeded supply. The project won the science and technology prize in a Japan/World Bank-sponsored network of environmentally beneficial development projects. The Kenyan experience is highly relevant to Asia.

Benefits in Asia

A number of biotechnology applications are widespread in Asia. These include biofertilizers and biopesticides, and products from tissue culture and micropropagation (banana, coconut, maize, potato, cassava). They also include diagnostics, vaccines, and embryo transfer for livestock. An ex-post evaluation on the impact of a transgenic crop at the farm level in Asia analyzed the impact of Bt cotton in the PRC. A sample of 283 cotton farmers from five counties of Hebei and Shandong provinces was studied. The key findings were (i) the cost to produce 1 kg of cotton was reduced by 20-30 percent, depending on the variety and site, and (ii) net income and returns to labor of all the Bt varieties were superior to the non-Bt varieties. Smaller farms and those farms that had lower incomes consistently derived larger increases in net income than larger farms and those with higher incomes.

The use of Bt cotton substantially reduced farmers' use of pesticides, usually from 12 sprays per season to 3 or 4. The study also found some preliminary evidence of positive impact on farmers' health. An ex-ante study on the impact of the introduction of biotechnology for the reduction of corn-borer infestation in the Philippines has postulated three kinds of benefits. First, the introduction of transgenic Bt maize would enhance the competitiveness of the Philippines from the point of view of import substitution and export prospects. Second, it would increase farmers' incomes. Third, it would reduce pesticide use.

WEIGHING RISKS AND BENEFITS OF BIOTECHNOLOGY

The evidence thus far available suggests potential economic, social, and environmental benefits from the use of biotechnology. It also Biotechnology to Address Poverty Reduction/Food Security Problems 61 suggests benefits for both food and export crops, for commercial and small farmers, and for consumers in the form of nutrition-enhanced products. The potential benefits will be realized only if a number of conditions are met:

- Public institutions will need to play a key role in directing the research agenda towards social and equity goals.
- Complementarity between the public and private sectors, in both R&D and in technology development and delivery, needs to be maximized so that public investment and research is focused on those areas not of interest to the private sector.
- Governments must be committed to developing technologies for the public good to ensure that the poor, including women, have access to the new technology.

The benefits and risks of biotechnology need to be assessed on a case-by-case basis. The uncertainties and the risks are yet to be fully understood, and the benefits are not yet fully exploited. It seems important not to deny people access to new biotechnology, so long as they are fully informed of the potential risks and benefits in making their choices. We have an ethical imperative not only to keep the technology portfolio open to biotechnology and genetic engineering, but also not to lose time. Every minute lost, every decision delayed, means more deaths from starvation and malnutrition.

POLICY AND PRIORITY SETTING

Issues

A surprising amount of biotechnology R&D is already being done in Asian developing countries. There have been few studies, however, on the cost-effectiveness of biotechnology R&D in these national programmes, and whether priorities are research-driven or demand-driven. On the other hand, countries have a justifiable concern that they will be left behind if not involved in the technology. In establishing policies and priorities, Asian countries should consider that (i) biotechnology requires a high level of capital and specialized technical skills, (ii) the public sector is constrained by inadequate capacity, (iii) the private sector has little interest in investing in technology for poor farmers, (iv) there are potential risks associated with GMO development, and (v) it is difficult to establish an effective biosafety framework. Some of the broad policy issues relate to capture of benefits. How can the benefits of biotechnology reach poor farmers? That is, how can biotechnology programmes be better linked to extension programmes? Who will be winners and losers from genetically engineered crops? Will there be a widening income gap between poor farmers and rich farmers who have access to credit to purchase biotechnology-derived seeds?

Options

There is a need to assist DMCs in policy and priority setting. Policy setting must be based on informed decisions on appropriate choices regarding the use of new biotechnology applications in agriculture and their prospects for enabling sustainable productivity. The likely impact of biotechnology on poor people, either directly by increasing crop yields and farm incomes or indirectly by improving their environment, should be considered. There is a need to encourage regional and national coordination of policy on biotechnology (especially of GMOs), and consistent distribution of responsibilities between agencies in different countries. That may be difficult to achieve, but the problems of poor coordination have been highlighted by the European experience. If Asian countries are to participate more fully in biotechnology, they will also need to expand their own capacities to undertake biotechnology research linked to the problems of smallholders and marginal farmers.

PRIORITIES FOR RESEARCH AND DEVELOPMENT

Issues

Countries have a range of options in determining policies and priorities for investment in biotechnology research. These might range from active promotion of biotechnology approaches, to a wait and see, stance, or to rejection of biotechnology R&D altogether. Whatever the case, technical and policy capacity will still be needed to support international trade and biosafety agreements. Donors, too, have a range of options. Should they focus their biotechnology investment on the more advanced developing countries, where there is more capacity and more likelihood of using the results? In any event, countries need to establish planning for investments based on demand. For example, is excessive emphasis given to plant biotechnology at the expense of animal biotechnology— especially given that the livestock sector is much faster growing?

There is a need to establish where biotechnology methods could deliver varieties that conventional breeding cannot. Where either a conventional or biotechnology-based breeding approach could be used to target a particular characteristic, there is the need for comparative economic and risk analyses of the alternative approaches. That should also take into account that many genetic engineering approaches also require conventional crossing and selection to transfer the desired character from the transformed plant into an adapted, elite variety. Many early biotechnology research programmes in Australia made the mistake of operating in isolation from breeding and agronomic improvement programmes, hindering application of the research. It is important that developing countries do not repeat the mistake.

Options

Obtaining a clear view of which crops and characters should be targets for genetic manipulation in Asian developing countries is a critical step, but it has rarely been addressed systematically. Research should focus on crops relevant to small farmers and poor consumers in developing countries:

cassava, yam, sweetpotato, rice, maize, wheat, millet, and possibly papaya. The limitation is that relatively little biotechnology research is being undertaken on many of Asia's basic food crops or on the problems of small farmers in rainfed and marginal lands. In many cases, Asian countries will need to expand their own national and regional capacity to undertake biotechnology research to address these problems. Once a clear view is obtained on which crops and characters to target, it may make more scientific and economic sense to purchase, license, or import particular elements of technology. Indeed, there have been a number of successes of crop genetic engineering in developing countries with technology already commercialized or in field trials. These include: insect resistance in maize and cotton and several horticultural crops using genes from Bt and other sources; resistance to a range of viruses in potato (viruses x, y, and leafroll); and ringspot virus resistance in papaya and cucurbits. Rice with resistance to bacterial blight and with increased content of beta-carotene or iron has been produced, as well as tomatoes with elevated lycopene (related to vitamin A).

Support of agricultural biotechnology research is important to developing countries for several reasons. Apart from direct benefits from applications, developing country ownership of IP can be traded for other technology. The existence of active programmes in-country can provide a balanced understanding of issues by technocrats who make policy in areas such as biosafety. Some other R&D needs include the following:

- One of the most profitable applications of crop biotechnology in the short term to medium term will be through better implementation of molecular markers for selection in conventional breeding programmes. That will require closer linkages between biotechnologists and plant breeders in many institutes.
- Engineering of desirable traits in several of the major cereal crops of Asia (*e.g.*, rice, maize, wheat) is being addressed by some of the international networks and through programmes at IARCs. Developing country NARSs need to strengthen their ability to adopt the technologies and the germplasm as they are developed by these international initiatives.
- It will be important to identify specific orphan crops and animals with short-to medium-term potential for application of existing biotechnology in developing countries. In some cases, this may best be done through developing more efficient methods of crop improvement or through the use of MAS for complex traits. In others, the most appropriate approach may be through the development of transgenic varieties with specific characteristics such as disease resistance or improved nutritional quality.
- There is a need to ensure that the genomic knowledge of the major food crops in Asia is obtained and is available in the public domain

as the basis for the development of new crop varieties for the public good. Such knowledge will underpin the development of the new food crop varieties for Asian agriculture over the next decades. The rice genomics programme would be a good model for others.

- Apomixis in a wider range of crops has huge potential for simplifying the breeding of adapted genotypes, preservation of hybrid vigor, and improved propagation of crops that currently rely on vegetative propagation. It is also an important tool for the discovery of genes that are potentially important in stress tolerance.
- There is a need to build the capacity of Asian developing countries to undertake laboratory and on-site testing of GMOs. An increasing number of published protocols are available as well as simple, commercial test kits.

INTELLECTUAL PROPERTY MANAGEMENT

Issues

Because the private sector invests heavily in, and holds many of the advanced biotechnologies, new discoveries in biotechnology may be protected by plant variety protection, patents, or trade secrets. This raises the issue of IPR. The 1995 Agreement on Trade-Related Aspects of Intellectual Property (TRIP) requires all countries to provide some sort of protection for plant varieties. TRIP requires all signatories to extend IP protection to microorganisms, plant genetic material, and techniques used for genetic manipulation. Although plants and animals, other than microorganisms, may be excluded from patent protection, countries are required to provide protection either by patents or by an effective sui generis system. Strengthened IPRs will increase the flow of technologies and products from developed to developing countries, and provide new incentives for local research and innovation. But one fear is that industrial nations, using genetic resources originating from developing nations, could develop GMOs or techniques and then restrict developing nations' access to the technology by employing the IPR provision of TRIP. Asian countries need to (i) develop a policy towards IP to protect their own discoveries, (ii) develop an effective approach to cooperation with the private sector in R&D, and (iii) encourage private sector research on products to benefit small farmers in Asia.

IP protection for plant genetic resources are being considered in the context of the Convention on Biological Diversity and renegotiation of the International Undertaking on Plant Genetic Resources of FAO. The issue has also been discussed in the World Intellectual Property Office in the context of traditional resource rights. In addition, environmental aspects of access to genetic resources are being discussed in UNEP and World Trade Organization (WTO). All these discussions are intended to ensure that existing international agreements encourage biotechnology development relevant to Asia's poor while protecting IPR of indigenous peoples. Most Asian countries are members of the World Intellectual Property Office, have a national patent regime, and are signatories to the Patent Cooperation Treaty and to the Paris Convention. With the exception of the PRC, whose application is pending, most are also members of WTO.

In conformity with TRIP, a number of countries are developing plant variety protection laws and are in the process of joining the International Convention for the Protection of Plant Varieties. These include India, Indonesia, Malaysia, Philippines and Viet Nam. Some countries (Bangladesh, India, Philippines) are currently formulating laws for the protection of biodiversity and community knowledge.

Proprietary interests in biotechnology affect developing countries in several ways, requiring them to consider issues relating to the (i) *licensing in* of technology owned by developed countries or multinational companies, (ii) their *freedom to operate* with their own inventions, and (iii) their ability to commercialize or export products. Changes in the last couple of decades in the IP environment in which agriculture operates include the patenting of living organisms in some countries, and establishment of plant variety rights. The protection of target genes and of enabling technologies (*e.g.*, gene regulation, markers for selection, promoter sequences, and transformation technologies) mean that developing country organizations have to be conversant with licensing activities.

Among them are technology use agreements, material transfer agreements, and commercial licenses. Some alternative strategies to gain freedom to operate can include: (i) inventing around current patents, (ii) redesigning constructs to synthesize genes to reduce reliance on external technical property, (iii) asking IP owners to relinquish claims or provide royalty-free licenses, (iv) ignoring all IP and technical property; or (v) seeking licenses for all IP and commercial property, which is certainly the safest route to building public-private sector cooperation. Many developing countries, however, use such *inputs* without formal permission.

This issue becomes important at the time of commercialization of a technique or process, or release of a variety. By then, it is often too late to change direction or incorporate an alternative piece of technology in a research or breeding programme. A recent ISNAR survey of five Latin American countries showed that NARS laboratories used proprietary technologies widely, but in many cases without formal agreements. Most of the proprietary technologies related to plant genetic engineering. They included markers for selection, transformation systems, promoters, and specific genes as well as marker techniques and diagnostics. There was also a lack of knowledge regarding IPRs in both academic and administrative ranks.

IP management of biotechnology inventions can be very complex. Kryder et al. (2000) recently reviewed the example of Golden Rice, which is high in pro-vitamin A (beta-carotene). This rice is potentially valuable for those who are too poor to obtain green vegetables. A very large number of proprietary technologies were used to develop Golden Rice. It is a multitransformant, since three genes were introduced into a carotenoid biosynthetic pathway to produce the high beta-carotene levels. Other proprietary technologies included three transformation vectors: *Agrobacterium* transformation, plant regeneration, and DNA amplification. Thus, a country or organization that requires this technology must obtain consent from a wide range of partners.

Options

The IPR issue is complex and yet many Asian institutes have focused training only on the technical aspects of biotechnology research, rather than on other areas such as licensing and IP management. Steps required to obtain consent from a wide range of partners for a given technology need to be defined on a case-by-case basis. Therefore, training in IP regulation to facilitate technology transfer in evolving IPR systems, and assistance in developing a clear understanding of TRIP agreements is important. Systems are needed to enable countries to protect their own technology, while minimizing barriers that could hinder technology transfer between countries The adoption of IPR law by developing countries may make proprietary biotechnologies (seeds and planting materials) more costly. To overcome this problem, the public sector would need to consider buying exclusive rights to newly developed technology and make it available free to small farmers. Alternatively, the public sector and IARCs could invest in biotechnology R&D so that the resulting technology could be made available free to farmers.

PUBLIC-PRIVATE SECTOR PARTNERSHIP

Issues

Significant increase in biotechnology investment by governments and donor agencies is crucial for achieving food security and poverty reduction. As mentioned earlier, there is insufficient profit motive to induce the private sector to undertake the R&D needed by small farmers. As Byerlee and Fischer (2000) point out, there have been significant market failures in applying biotechnology for the benefit of developing country farmers; therefore the public sector will continue to have an essential role. Because many of the techniques and products of modern biotechnology are privately owned, public agencies need modes of action to operate in an increasingly private sector world. Furthermore, both the public and private sectors have complementary assets needed for biotechnology to be applied to its full potential.

Many of the results of biotechnology research will be most easily transferred to poor farmers as seeds in much the same way as the results of the Green Revolution. But this time it is expected that the private sector will have a significant role in dissemination. The private sector, especially seed companies, is becoming more important in several Asian developing countries such as India and the PRC. Similarly, in Europe and North America several seed companies have formed alliances with global life science companies. Several Asian agrochemical and floriculture companies are involved in joint ventures with developed country companies. Thus, in line with ADB's Private Sector Development Strategy (ADB 2000), there is a need to increase the partnership between the public and private sectors.

Options

Present involvement of companies in agricultural biotechnology in developing countries in Asia (especially in major crops and large countries) suggests that the motivation is purely commercial. For example, seed for up to half the engineered cotton grown in the PRC has been commercially obtained from Monsanto. In many cases new mechanisms will be needed by which the private sector can assist with technology transfer, extension, and distribution of biotechnology products.

That may require some innovative arrangements involving special funds and other financial incentives for private companies. Multinationals also have an incentive to involve themselves in developing country partnerships/ philanthropy to counter negative public impressions. Segmentation of markets will be important to allow developing country farmers to access the products of biotechnology under realistic conditions.

Brokering groups such as ISAAA, a USAID programme, and CGIAR centers have already implemented a range of approaches. These include market segmentation based on crop and growth region, country income level, trade status, or crop variety. In other cases the public sector could offer to buy exclusive rights to newly developed technology and make it available either free or for a nominal charge to small farmers (but on a profit basis to commercial farmers). The private research agency would bear the risks, as it does when developing technology for the market. The other possibility is for the public sector agency to finance private R&D on orphan crops through competitive bidding.

There is a need to establish whether international organizations and funding agencies should play a brokering role in the dissemination of agricultural biotechnology. A neutral broker can have advantages. For example, ISAAA claims that its comparative advantages are cosponsorship by public and private sector institutions, independence, and neutrality (lack of financial interest in the technology). Funding agencies (*e.g.*, USAID) have also served as brokers between public agencies and private companies. Centralization of technology transfer offices for agricultural biotechnology in individual countries would facilitate brokering agreements, and inspire greater confidence in the private sector.

Other biotechnologies such as micropropagation of plants through tissue culture, and production of biofertilizers, biopesticides, and selected vaccines may be suitable for small-and medium-sized private companies in developing countries. Such companies could take technology directly from local research institutes. There are many successful examples of this already. Such enterprises can create employment in rural areas and deliver affordable and useful products to farmers.

BIOSAFETY

Issues

Biosafety concerns relate either to food safety and human health or to the environmental impact of genetically modified crops. There is also a broader ethical concern as to whether genetically modified foods are unnatural, and whether the use of the technology encourages the narrow control of agriculture by a few (multinational) corporations. One of the key principles in biosafety guidelines is that of *substantial equivalence* as the approach to identify differences between biotechnology-derived and traditional foods. The main environmental risks relate to the (i) potential loss of genetic diversity in cropping systems; (ii) potential transfer of genes from herbicide-resistant crops to wild relatives, creating superweeds; (iii) ability of herbicide-resistant crops to act as weeds in rotation crops; (iv) escape of transgenes, especially antibiotic resistance markers to soil bacteria; (v) vector recombination to create new viruses; and (vi) with Bt toxins, insect resistance to the toxin and the effect of the toxin on nontarget organisms.

The need for developing countries to have functioning biosafety systems has strengthened since the adoption of the Cartagena Protocol on Biosafety in January 2000. The Protocol establishes a framework for regulating international trade in transgenic crops. Three major components of the Protocol have implications for individual countries as users, developers, and exporters of GMOs:

- International shipments that "may contain" transgenic food products must be so labeled. This applies only to large-scale shipments; it does not affect labeling requirements on consumer products, which are determined by each country.
- Governments may use the *precautionary principle* to bar import of a transgenic product even in the absence of conclusive evidence that the product is not safe. The Protocol does not, however, override other international agreements, including WTO, which requires that import decisions be sciencebased.
- To assist countries in making import decisions, a database will be established to make available uniform information on transgenic crop varieties.

The need for harmonization of regulations is not just an issue for developing economies in Asia, but also for developed countries. For example, Levidow et al. (1999) report on how difficult it has been for European countries to harmonize regulatory criteria despite an expressed position of using science-based criteria. Countries disagree on the amount of scientific evidence required to resolve uncertainties such as in the recent approval processes for Bt insect-resistant maize and herbicidetolerant canola.

Options

Countries have a range of options in designing and implementing national biosafety systems geared to their technical, legal, and institutional realities and in adapting the system to local needs, priorities, and capacities. There is a need to assess the adequacy and effectiveness of biosafety procedures for the testing, release, import, production, and use of GMOs in Asian countries.

Regulatory systems must be flexible to allow for either increased scrutiny or relaxation of controls (*e.g.*, for containment of field trials) based on available scientific evidence. Harmonization of regulations between agencies is important.

But it may be simpler and less expensive to embed biosafety regulation within existing institutions rather than to build new ones. Australia's approach has been to develop national policies and regulations for trade in GMOs within the country's existing regulatory framework. Thus, governments could be helped to build the infrastructure to manage GMOs within the context of developing a quarantine policy, managing sanitary and phytosanitary issues, and assessing risk and environmental impact. Collaboration between quarantine/regulatory officials and environmental policymakers should be strengthened.

An area of potential concern is the lack of verified ecological data on the effects of genetically modified crops under Asian or tropical conditions. Governments and international agencies should be supported in participatory field studies on the ecological impact of first-generation of genetically modified crops such as insect-resistant cotton and rice. These assessments should involve local communities in the evaluation of the new technologies.

The likelihood that Asian developing countries and their agricultural export customers will adopt the labeling of food containing genetically modified products needs to be assessed. Several countries have introduced labeling systems or are considering them. In some cases, industry has expressed concerns about the extra costs involved in testing, ingredient tracing, and labeling.

Several options for labeling have been proposed. They include (i) labeling based on the presence of detectable transgenic DNA or protein only; (ii) special labeling for GMO-free foods; (iii) labeling all foods derived from GMOs; and (iv) labeling foods or any food ingredients produced with GMOs (*e.g.*, meat from animals fed with transgenic crop residues). It will be important to establish the impact of such labeling on the production, distribution, marketing, and exports of genetically modified foodstuffs.

FOSTERING BETTER PUBLIC AWARENESS

Issues

The future of biotechnology lies in public awareness and acceptance; good technology alone is not enough. For example, although nuclear power provided low-cost electricity for many countries, there has been substantial disadoption in recent years. The news media in several Asian developing countries regularly report on biotechnology and the controversy surrounding GMOs. Some aspects of the debate are similar to 74 Agricultural Biotechnology, Poverty Reduction and Food Security that in Australia, in which the controversy is seen as one of technocrats encouraging the use of biotechnology versus the average person who is unsure of the benefits. European countries, especially, are seen to have the luxury of excess food production. The need to alleviate poverty or enhance food security is not part of their debate. Certain additional issues appear in some developing countries. Some countries such as Thailand are major exporters of processed foods. They are concerned about the attitudes of their export customers towards genetically modified foods.

Options

Inclusion of public awareness activities from the onset of a biotechnology programme, rather than at its completion, can greatly assist in gaining acceptance. For example, before starting development of a transgenic, insect-resistant, tropical maize in Kenya, CIMMYT consulted with NGOs, farmer organizations, media, and other stakeholders to create a supportive environment. Public awareness programmes should be included in the activities of major institutions carrying out biotechnology R&D. That is already done in Thailand and the Philippines, using simple messages in local languages. Again, it will be equally important to involve community groups in the debate, not just scientists or the food industry. Finally, multinational organizations such as Monsanto have recently realized the necessity for developing a strategy to ensure informed public debate about the risks of biotechnology to human health and the environment, and to ensure consumers are fully aware of the true benefits and costs of biotechnology. Such campaigns must recognize that freedom of choice is valid.

THE ROLE OF TRADE IN ENHANCING FOOD SECURITY

The definition of food security provided at the World Food Summit in 1996 describes the vision, but not necessarily the way to get there. Agricultural trade is seen by many as a key element in securing both availability of and access to food. The Summit declared that "trade is a key element in achieving food security. We agree to pursue food trade and overall trade policies that will encourage our producers and consumers to utilise available resources in an economically sound and sustainable manner".

This paper focuses on the role of agricultural trade in enhancing or undermining food security. It puts forward that trade should be less important to other activities in building food security. But insofar as there will be trade in food, recommendations are made seeking to eliminate the negative effects and enhance any potential positive effects of trade on food security.

GLOBAL FOOD CONSUMPTION

Most food produced in the world is consumed in the country of production as seen with the following percentages:

- Wheat is the largest export cereal crop, yet just 17% of total wheat production is traded internationally.
- Beef is the largest exported animal crop with just 10% of total production being traded.
- Soya beans, around which much of the global feed industry revolves, has an export share of 30% of total soya production.

However, these figures indicate a sharp increase from two decades previously.

THE THEORY OF INTERNATIONAL TRADE

Theoretically, free international trade allows the supply of goods to where they are demanded at least cost. The theory of comparative advantage holds that a country with a comparative advantage in a certain product will benefit by exporting that product and importing those products in which it does not have a comparative advantage. International trade in food allows countries unable to produce for themselves to buy from other countries at the lowest cost.

It also allows countries to produce something of greater value for export, and use some of the money earned to import products for which their opportunity cost of production is greater. For example, it may be economically more advantageous to build a factory on a piece of land and export the products, than to use the same piece of land to produce food crops that could be imported for less than the income received from the manufactured goods.

THE REALITY OF INTERNATIONAL AGRICULTURAL TRADE

The theory of comparative advantage would thus predict that the industrialised countries should produce and export processed and manufactured goods. Raw materials including agricultural products would therefore be imported. However, industrialised countries are the major exporters of most traded agricultural products.

Industrialised countries dominate exports of the following:

- wheat

- coarse grains
- soybeans
- meat and
- dairy products.

Industrial countries are also the major importers of agricultural products, with the exception of wheat and coarse grains.

Developing countries therefore import staples and export other agricultural products. Overall, industrialised countries dominate international trade in food, and they account for a greater percentage of exports than the theory of comparative advantage would predict. One reason for this is that most developed countries protect their agricultural industries. More detail on this will be presented later, in the discussion on the World Trade Organisation (WTO) and the EU's Common Agricultural Policy (CAP).

International agricultural trade thus does not work as the theory predicts it should, and there are a number of reasons for this.

The ability to influence world market prices: World prices should reflect the balance of world demand and supply. However, there are a number of barriers to this. For some products, individual countries have such a large number of buyers or sellers that they can have a disproportionate effect on world market prices. This is related to tendencies towards monopolisation or "cartelisation" of branches of agriculture. While consumers should benefit from cheaper imports, they may stand to lose in the long run if trade liberalisation gives more power to monopolies.

This refers to the influence of corporate hegemony – the "unmentioned agenda" of the ownership, control and concentration of the food and agricultural industry. This concentration of ownership and control leads to distortions in the structure and price mechanism of the world market. An important aspect of this is "planned scarcity" to stabilise food prices at the highest possible level over a prolonged period, thus allowing multinationals to accrue the largest profits possible.

Subsidisation of production in industrial countries: While it is assumed food commodities will be sold on world markets at the price of production, industrial countries subsidise production to undermine this.

The result of subsidised oversupply is to drag the world market price below the costs of even the most efficient producers. Government policies and resource transfers to producers gives industrialised countries the "competitive advantage". This means that developing country producers are unable to compete regardless of their efficiency. Therefore, available resources to provide support to producers become critical in determining whether a country will be a net exporter or a net importer of agricultural products.

International trade in agriculture is problematic in terms of pricing. Exports make up only a small proportion of most food and agricultural commodities, as discussed above. Countries usually produce first for their domestic markets and then export what is left. Thus, in years of shortage they will export less than normal, and in years of surplus they will export more than normal. If there is a variation of 5% from the norm year to year in a market where only 10% of the product is exported, this can result in a 50% increase or decrease in export volumes from year to year.

Since food and agricultural products have a low price elasticity of demand, prices will be highly volatile. This means that a small drop in the volume being traded can result in a rapid increase in the price, and that a small rise in volumes traded can result in a rapid drop in the price.

CAUSES OF DEBT IN DEVELOPING COUNTRIES

This volatility creates problems food importing countries, since they cannot be sure of a stable supply or how much the food will cost them. As noted earlier, developing countries are the main importers of wheat and coarse grain – staple crops. In fact, one direct causes of the current debt of developing countries is the conversion of the agricultural economies from self-sufficiency in staples to becoming food-importing countries. This conversion took place in the post World War 2 period combined with the wheat crisis in the early 1970s. In turn, the indebtedness has led to structural adjustment and underdevelopment.

The Theory of Free Trade

Free trade is based on the principle of supply and demand. Accordingly, a demand for a commodity is apparently best met through the mechanism of the free market not controlled by government intervention.

Effective Demand

However, the demand itself is always qualified when dealing with the market, since the market only responds to "effective demand", *i.e.* the ability to pay for the commodity at prevailing prices. This 'effective demand' excludes most of the world's population that lacks a "market-based entitlement" and is unable to purchase food as a commodity. Indeed, the market system is more likely to remove food from areas where the needs for it are greatest, since "effective demand" comes from those able to pay for the food, and not those unable to pay. Therefore, using current economic mechanisms, trade fails to distribute food to where it is really needed. Since the market is based on ownership rights, legal exchanges and so on, "the law stands between food availability and food entitlement".

From a production point of view, theoretical comparative advantage and incentives for higher production are fairly meaningless when the conditions for increased production are not present, and the resources necessary for production are inaccessible. International trade in food may end up only benefiting those with access to productive resources at scale, and those who are in the market to purchase food. Therefore, trade liberalisation may – and does – result in greater power to larger producers including multinational corporations, at the expense of small and resource-poor producers and consumers.

Food security requires the sustainable use of productive resources, because unsustainable use means that production will decline in the long term. From an ecological perspective, sustainable agriculture rests on the preservation of the productive capacity of natural systems, and the minimisation of non-renewable resources.

However, there is a growing contradiction between trade and sustainable resource use. It is increasingly difficult to meet the requirements of sustainable agriculture the further physically removed food production and consumption are from one another. This is because to retain high production levels with minimal use of non-renewable resources, agriculture must be based on optimal use of locally available resources. Large-scale production based on regional specialisation and long distance transport undermines this possibility.

There are many consequences of international trade in food as it happens in practice. It has an influence on both exports and imports. Since "effective demand" dictates what should be produced, production is skewed towards the requirements of those with resources.

Simply put, producers in developing countries are more likely to produce cash crops for export to lucrative markets in industrialised countries (or even domestically) than produce staples. Government subsidisation of producers in the industrialised countries reduces the world price. This has a negative effect on producers in the countries without subsidies, since developing countries have to compete against subsidised products. At the same time, subsidised exports from industrialised countries results in the dumping of surpluses in developing countries, mainly to the long-term detriment of local primary as well as secondary (processing) production.

THE RELATIONSHIP BETWEEN FOOD SECURITY AND INTERNATIONAL AGRICULTURAL TRADE

Many questions therefore need to be asked concerning the relationship between food security and international agricultural trade:

- What are the underlying reasons for trading in food?
- Is it to generate foreign exchange, or to ensure that all people have enough food to eat?
- Who leads the agenda and for what motives?

As it stands at present, trade policy in every country generally reflects commercial interests. It is not countries that trade, but companies, and this suggests that trade negotiations are carried out by governments on behalf of particular corporations. Overall, a distinction needs to be made between food security concerns and agricultural export concerns. The "right to export" that underpins trade agreements is not compatible with food security. In recent times, there has been an increasing alignment of international agreements. It has become impossible for international negotiations to be held without reference to other discussions and agreements made elsewhere by the same parties. The World Summit on Sustainable Development (WSSD) is one in a line of upcoming meetings of governments to deal with issues of global scope.

The power of the WTO, and its position as the only enforceable agreement, has established trade as the over-riding priority for international agreements. The role of trade in enhancing or undermining sustainable agriculture, and hence food security, will therefore have to be dealt with at the WSSD.

Group for Environmental Monitoring's position

Group for Environmental Monitoring's (GEM) position on engaging with international fora is to "strive for equity and environmental justice by contributing to the formulation and implementation of integrated local, national and international policy.

There is a philosophy to use resource economics tools to influence the national and global policy within the existing "conventional wisdom" paradigm while at the same time attempting to "shift that paradigm". From this perspective, the issue is how to use positions within the formal framework to strengthen the ability and space of people to develop alternatives parallel to or outside the formal framework. This can be achieved not only by creating legal space, but also by allowing people to begin to set up practical alternatives that will eventually supersede the top-down and corporate driven globalisation of the present. With this approach in mind, key issues relating to the global processes will be spelled out, followed by the South African government's positions in these processes and a consideration of the potential impacts on food security of these positions. Recommendations on positions in the global processes, with specific regard to agricultural trade, will then be provided.

This will be followed by a consideration of how these recommended positions may contribute to creating the necessary alignment with forces seeking to shift the "conventional paradigm".

FOOD SECURITY AND STARVATION

Food security describes a situation in which people do not live in hunger or fear of starvation. World-wide around 852 million men, women and children are chronically hungry due to extreme poverty; while up to 2 billion people lack food security intermittently due to varying degrees of poverty.

A direct relationship exists between food consumption levels and poverty. Families with the financial resources to escape extreme poverty rarely suffer from chronic hunger; while poor families not only suffer the most from chronic hunger, but are also the segment of the population most at risk during food shortages and famines.

Two commonly used definitions of food security come from the UN's Food and Agriculture Organization (FAO) and the United States Department of Agriculture (USDA):

- Food security exists when all people, at all times, have access to sufficient, safe and nutritious food to meet their dietary needs and food preferences for an active and healthy life.
- Food security for a household means access by all members at all times to enough food for an active, healthy life. Food security includes at a minimum
  * The ready availability of nutritionally adequate and safe foods,
  * An assured ability to acquire acceptable foods in socially acceptable ways (that is, without resorting to emergency food supplies, scavenging, stealing, or other coping strategies).

The stages of food insecurity range from food secure situations to full-scale famine. "Famine and hunger are both rooted in food insecurity. Food insecurity can be categorized as either chronic or transitory. Chronic food insecurity translates into a high degree of vulnerability to famine and hunger; ensuring food security presupposes elimination of that vulnerability. Chronic hunger is not famine. It is similar to undernourishment and is related to poverty. It exists mainly in poor countries."

FOOD INSECURITY

Food insecurity has been described as "a condition in which people lack basic food intake to provide them with the energy and nutrients for fully productive lives." Hunger Task Force

Food insecurity is defined as not having access to enough food every day of the year. The U.S. federal government collects statistics on this based on surveys. Thus a "food insecure household" is one in which any member experienced one or more days in a given year when they didn't have enough food because of poverty or a lack of resources. This poverty measure takes into account participation in food assistance programmes such as school breakfast, school lunch and Food Stamps.

The USDA report cited below asks the question, "How often were people hungry in households that were food insecure with hunger?" Around 4 percent of people reported going hungry at least once a year, while on any given day the figure is estimated to be between 0.5 percent and 0.8 percent.

Chronic Nutritional Deficiencies

Many countries experience perpetual food shortages and distribution problems. These result in chronic and often widespread hunger amongst significant numbers of people. Human populations respond to chronic hunger and malnutrition by decreasing body size, known in medical terms as stunting or stunted growth. This process starts *in utero* if the mother is malnourished and continues through approximately the third year of life. It leads to higher infant and child mortality, but at rates far lower than during famines. Once stunting has occurred, improved nutritional intake later in life cannot reverse the damage. Stunting itself is viewed as a coping mechanism, designed to bring body size into alignment with the calories available during adulthood in the location where the child is born. Limiting body size as a way of adapting to low levels of energy (calories) adversely affects health in three ways:

- Premature failure of vital organs occurs during adulthood. For example a 50 year old individual might die of heart failure because his/her heart suffered structural defects during early development.
- Stunted individuals suffer a far higher rate of disease and illness than those who have not undergone stunting.
- Severe malnutrition in early childhood often leads to defects in cognitive development.

"The analysis points to the misleading nature of the concept of subsistence as Malthus originally used it and as it is still widely used today. Subsistence in not located at the edge of a nutritional cliff, beyond which lies demographic disaster. Rather than one level of subsistence, there are numerous levels at which a population and a food supply can be in equilibrium in the sense that they can be indefinitely sustained. However, some levels will have smaller people and higher normal mortality than others."

Dictatorship and Kleptocracy

As the Nobel Prize-winning economist Amartya Sen has observed that "there is no such thing as an apolitical food problem." While drought and other naturally occurring events may trigger famine conditions, it is government action or inaction that determines its severity, and often even whether or not a famine will occur. The 20th century is full of examples of governments undermining the food security of their own nations – sometimes intentionally.

Socialist anti-Western governments do this much more than authoritarian, pro-Western ones. Prime examples are North Korea since 1980, and the Soviet Union in the mid-20th century. Lesser known are Cambodia under Pol Pot, and the Sandinistas in Nicaragua in their attempt to starve the Misquito Indians into submission.

When governments come to power by force or rigged elections, and not by way of fair and open elections, their base of support is often narrow and built upon cronyism and patronage. Under such conditions "The distribution of food within a country is a political issue. Governments in most countries give priority to urban areas, since that is where the most influential and powerful families and enterprises are usually located. The government often neglects subsistence farmers and rural areas in general. The more remote and underdeveloped the area the less likely the government will be to effectively meet its needs. Many agrarian policies, especially the pricing of agricultural commodities, discriminate against rural areas. Governments often keep prices of basic grains at such artificially low levels that subsistence producers can not accumulate enough capital to make investments to improve their production. Thus, they are effectively prevented from getting out of their precarious situation."

Further Communist parties, as well as dictators and warlords have used food as a political weapon, rewarding their supporters while denying food supplies to areas that oppose their rule. Under such conditions food becomes a currency with which to buy support and famine becomes an effective weapon to be used against the opposition.

Governments with strong tendencies towards kleptocracy can undermine food security even when harvests are good. When government monopolizes trade, farmers may find that they are free to grow cash crops for export, but under penalty of law only able to sell their crops to government buyers at prices far below the world market price. The government then is free to sell their crop on the world market at full price, pocketing the difference. This creates an artificial "poverty trap" from which even the most hard working and motivated farmers may not escape.

When the rule of law is absent, or private property is non-existent, farmers have little incentive to improve their productivity. If a farm becomes noticeably more productive than neighbouring farms, it may become the target of individuals well connected to the government. Rather than risk being noticed and possibly losing their land, farmers may be content with the perceived safety of mediocrity. As pointed out by William Bernstein in his book *The Birth of Plenty*: "Individuals without property are susceptible to starvation, and it is much easier to bend the fearful and hungry to the will of the state. If a farmer's property can be arbitrarily threatened by the state, that power will inevitably be employed to intimidate those with divergent political and religious opinions."

Economic Approaches

There are many economic approaches advocated to improve food security in developing countries. Three typical approaches are listed below. The first is typical of what is advocated by most governments and international agencies. The other two are more common to non-governmental organizations (NGO's).

Westernized View

Conventional thinking in westernized countries is that maximizing the farmers profit is the surest way of maximizing agricultural production; the higher a farmer's profit, the greater the effort that will be forthcoming, and the greater the risk the farmer is willing to take.

This view holds that it is the governments job to place into the hands of farmers the largest number and highest quality tools possible (tools is used here to refer to improved production techniques, improved seeds, secure land tenure, accurate weather forecasts, etc.) However, it is left to the individual farmer to pick and choose which tools to use, and how to use them, as farmers have intimate knowledge of their own land and local conditions.

As with other businesses, a percentage of the profits are normally reinvested into the business in the hopes of increasing production, and hence increase future profits. Normally higher profits translate into higher spending on technologies designed to boost production, such as drip irrigation systems, agriculture education, and greenhouses. An increased profit also increases the farmer's incentive to engage in double-cropping, soil improvement programmes, and expanding usable acreage.

Food Justice

An alternative view takes a collective approach to achieve food security. It notes that globally enough food is produced to feed the entire world population at a level adequate to ensure that everyone can be free of hunger and fear of starvation. That no one should live without enough food because of economic constraints or social inequalities is the basic goal.

This approach is often referred to as food justice and views food security as a basic human right. It advocates fairer distribution of food, particularly grain crops, as a means of ending chronic hunger and malnutrition. The core of the Food Justice movement is the belief that what is lacking is not food, but the political will to fairly distribute food regardless of the recipient's ability to pay.

Food Sovereignty

A third approach is known as food sovereignty; though it overlaps with food justice on several points, the two are not identical. It views the business practices of multinational corporations as a form of neocolonialism. It contends that multinational corporations have the financial resources available to buy up the agricultural resources of impoverished nations, particularly in the tropics. They also have the political clout to convert these resources to the exclusive production of cash crops for sale to industrialized nations outside of the tropics, and in the process to squeeze the poor off of the more productive lands. Under this view subsistence farmers are left to cultivate only lands that are so marginal in terms of productivity as to be of no interest to the multinational corporations.

It advocates banning the production of most cash crops in developing nations, thereby leaving the local farmers to concentrate on subsistence crops.

In addition it opposes allowing low-cost subsidized food from industrialized nations into developing countries, what is referred to as "import dumping".

World Food Summit

The World Food Summit was held in Rome in 1996, with the aim of renewing global commitment to the fight against hunger. The Food and Agriculture Organization of the United Nations (FAO) called the summit in response to widespread under-nutrition and growing concern about the capacity of agriculture to meet future food needs. The conference produced two key documents, the Rome Declaration on World Food Security and the World Food Summit Plan of Action.

The Rome Declaration calls for the members of the United Nations to work to halve the number of chronically undernourished people on the Earth by the year 2015. The Plan of Action sets a number of targets for government and non-governmental organizations for achieving food security, at the individual, household, national, regional and global levels.

ACHIEVING FOOD SECURITY

"The number of people without enough food to eat on a regular basis remains stubbornly high, at over 800 million, and is not falling significantly. Over 60 per cent of the world's undernourished people live in Asia, and a quarter in Africa. The proportion of people who are hungry, however, is greater in Africa (33 per cent) than Asia (16 per cent). The latest FAO figures indicate that there are 22 countries, 16 of which are in Africa, in which the undernourishment prevalence rate is over 35 per cent." – Food and Agriculture Organization (FAO)

'In general the countries that succeeded in reducing hunger were characterised by more rapid economic growth and specifically more rapid growth in their agricultural sectors. They also exhibited slower population growth, lower levels of HIV and higher ranking in the Human Development Index'.

USAID proposes several key steps to increasing agricultural productivity which is in turn key to increasing rural income and reducing food insecurity. They include:

- Boosting agricultural science and technology. Current agricultural yields are insufficient to feed the growing populations. Eventually, the rising agricultural productivity drives economic growth.
- Securing property rights and access to finance.
- Enhancing human capital through education and improved health.
- Conflict prevention and resolution mechanisms and democracy and governance based on principles of accountability and transparency in public institutions and the rule of law are basic to reducing vulnerable members of society.

The First Millennium Development Goal is to Eradicate extreme hunger and poverty, and agricultural productivity is likely to play a key role if it is to be reached on time.

"Of the eight Millennium Development Goals, eradicating extreme hunger and poverty depends on agriculture the most.

The Agriculture-hunger-poverty Nexus

Eradicating hunger and poverty requires an understanding of the ways in which these two injustices interconnect. Hunger, and the malnourishment that accompanies it, prevents poor people from escaping poverty because it diminishes their ability to learn, work, and care for themselves and their family members. Food insecurity exists when people are undernourished as a result of the physical unavailability of food, their lack of social or economic access to adequate food, and/or inadequate food utilization. Food-insecure people are those individuals whose food intake falls below their minimum calorie (energy) requirements, as well as those who exhibit physical symptoms caused by energy and nutrient deficiencies resulting from an inadequate or unbalanced diet or from the body's inability to use food effectively because of infection or disease.

An alternative view would define the concept of food insecurity as referring only to the consequence of inadequate consumption of nutritious food, considering the physiological utilization of food by the body as being within the domain of nutrition and health.

Malnourishment also leads to poor health hence individuals fail to provide for their families. If left unaddressed, hunger sets in motion an array of outcomes that perpetuate malnutrition, reduce the ability of adults to work and to give birth to healthy children, and erode children's ability to learn and lead productive, healthy, and happy lives. This truncation of human development undermines a country's potential for economic development – for generations to come.

There are strong, direct relationships between agricultural productivity, hunger, and poverty. Three-quarters of the world's poor live in rural areas and make their living from agriculture. Hunger and child malnutrition are greater in these areas than in urban areas. Moreover, the higher the proportion of the rural population that obtains its income solely from subsistence farming (without the benefit of pro-poor technologies and access to markets), the higher the incidence of malnutrition. Therefore, improvements in agricultural productivity aimed at small-scale farmers will benefit the rural poor first.

Increased agricultural productivity enables farmers to grow more food, which translates into better diets and, under market conditions that offer a level playing field, into higher farm incomes. With more money, farmers are more likely to diversify production and grow higher-value crops, benefiting not only themselves but the economy as a whole."

Biotechnology for Smallholders

The area sown to genetically enhanced crops in developing countries is rapidly catching-up with the area sown in industrial nations. According to the International Service for the Acquisition of Agri-biotech Applications (ISAAA), genetically enhanced (biotech, GM) crops were grown by approximately 8.5 million farmers in 21 countries in 2005, up from 8.25 million farmers in 17 countries in 2004. The largest increase in biotech crop area in any country in 2005 was in Brazil, provisionally estimated at 44,000 km^2 (94,000 km^2 in 2005 compared with 50,000 km^2 in 2004. India had by far the largest year-on-year proportional increase, with almost a threefold increase from 5,000 km^2 in 2004 to 13,000 km^2 in 2005. Current high regulatory costs imposed on varieties created by the more modern methods are a significant hurdle for development of genetically enhanced crops well suited to developing country farmers by modern genetic methods. Once a new variety is developed, however, seed provides a good vehicle for distribution of improvements in a package that is familiar to the farmer. Currently there are some institutes and research groups that have projects in which biotechnology is shared with contact people in less-developed countries on a non-profit basis. These institutes make use of biotechnological methods that do not involve high research and registration costs, such as conservation and multiplication of germplasm and phytosanitation.

CAUSES AND BACKGROUND OF FOOD SECURITY

CHANGES IN THE FOOD SYSTEM

The Indian food supply is abundant and affordable and it is acknowledged by many to pose an acceptable level of risk. The food system has evolved from one that provided consumers with minimally processed basic commodities for home meal preparation to today's availability of highly processed products that are ready-to-eat or require minimal preparation. Food preservation processes have changed dramatically from traditional salting, curing, drying, and heating. Today's products are the result of many technological developments such as pasteurization, irradiation, and genetic engineering. Likewise, food distribution systems have changed greatly. The broad introduction of refrigerated railcars and trucks, freezers, and air transport created a national and now global food system. These changes allow out-of-season availability, convenience, variety, and improved sensory attributes. The size and complexity of the Indian food system raise many safety issues, for example:

- The emergence of new foodborne pathogens such as *Escherichia coli* 0157:H7 and the re-emergence of previously identified pathogens such *as Salmonella* have resulted in new microbiological hazards.

- Advances in science and technology that allow the development of genetically modified foods and the construction of modified macronutrients require new ways of evaluating the safety of substances added to the food supply, and this need will increase.
- Heightened consumer interest in raw or minimally processed fruits and vegetables, partly in response to dietary recommendations, has created a year-round demand for fresh produce, which can be met out-of-season only through increased imports. This increased produce volume requires additional attention to possible contamination of domestic as well as imported fruits and vegetables.
- Americans eat fewer home-prepared meals than ever before, in response to changes in the US workforce and to developments in food processing and food service that offer greater convenience and variety in available foods. The potential for contamination increases as food is handled by more people.
- As technology has advanced, a smaller number of food facilities provide food to larger numbers of Indian consumers, increasing the extent of harm that can arise from a single incident.
- The remarkable success of modern medicine in extending the lifespan and increasing the quality of life for many people has placed new demands on the food system and on those responsible for guarding its safety. Increasing numbers of people have immune systems that are compromised because of age, illness, or medical treatment. These people are highly susceptible to illness and death from microbial pathogens, and might be more sensitive to new food ingredients and recently identified natural components of the diet.
- Increasing consumption of fortified foods and dietary supplements, including herbals, has raised new questions about the safety of "natural" substances not normally in the diet, or normally part of the diet but at much lower concentrations, and about the health effects of consuming high concentrations of nutrients ordinarily considered safe.

Thus, the developments that have provided the American consumer with a wide array of food products have also introduced risks. Government has attempted to address such developments by adding structures and processes without always considering their effects on other aspects of the system. As a result, inefficiencies and apparent conflicts within the system have arisen. Some have been corrected (for example, pesticides have been exempted from the Delaney clause's ban on carcinogens), but others remain. For example, inspectors from multiple agencies oversee parallel and nearly identical processes in the same food processing facility. The current food safety system has evolved piecemeal over almost a century in response to changes in the food supply and to changes in the biophysical and social environments in which the system operates. The present system is not the product of planning, and it is often not equipped to anticipate changes. But the situation is not just haphazard; changes in risks have made the system outmoded. The role and organization of government entities have remained largely unchanged, and the food safety system has fallen behind today's needs.

Scope of the Food Safety Problem

Over the years, as the agriculture and marketing systems have evolved to provide food to a growing and increasingly sophisticated population, complex processes built on advances in science and technology have been developed to evaluate and manage the risks associated with the food supply. In spite of well-established systems that control many food risks, serious hazards to public health remain. Although estimates vary greatly, there is agreement that foodborne illness is a serious problem. In the United States, an estimated 81 million cases of foodborne illness may occur each year, resulting in as many as 9,000 deaths. The estimated annual medical costs and productivity losses due to seven major foodborne pathogens range from $6.6 billion to $37.1 billion.

The responsibility for managing foodborne risks is shared throughout the system because the wholesomeness and safety of a food are influenced by all the people and processes that handle or transform it from production to consumption. Federal, state, and local governments play major roles in managing risks to protect the public from hazards in the food supply. Regulatory agencies are empowered to prescribe rules, standards, and processes to control risks; to develop and maintain research programmes to apply contemporary science and technology to safety decisions; to monitor risks in the food supply; and to provide information and education to all components of the food system. In authorizing and funding efforts to ensure the safety of the food supply, the government must balance the interests of diverse groups and allocate finite resources among competing needs.

Faulty handling of a product at any point in the system can transform a safe product into one that can cause serious harm. Producers, shippers, importers, processors, wholesalers, retailers, handlers, and consumers all influence the health risks associated with food products. Consumers play a particularly important role in the control of microbiological risks, both in their food handling practices and in their demand for an effective, efficient food safety system.

Food safety issues also involve the interplay of domestic and international legal, political, scientific, social, and economic forces. Intense debates about desired levels of protection and about the appropriateness of different control measures can arise as parties discuss the scientific bases for risk decisions, public expectations, and relative costs and benefits of intervention in different ways and in different components of the food system.

Science must play a vital role in food safety decisions through risk assessment, that is, the identification of hazards and the determination of the likelihood and severity of risk under given conditions of exposure. Hazards are biological, chemical, or physical substances that can cause adverse consequences. Hazards associated with food include microbiological pathogens, naturally occurring toxins, allergens, intentional and unintentional additives, modified food components, agricultural chemicals, environmental contaminants, animal drug residues, and excessive consumption of some dietary supplements. In addition, certain methods of food preparation in the home can contribute to increasing some of these hazards. However, a hazard does not pose a risk in the absence of exposure. The susceptibility of the consumer and the magnitude of exposure determine whether those hazards cause immunological changes, genetic and developmental changes, cancer, or death.

Hazard identification is the basis for estimating risk. A food safety risk is the probability of harm to health resulting from a food-related hazard at a particular exposure to a specified person or group. It is important to recognize that safety is an intellectual concept, not an inherent biological property of a substance; safety has been defined as the judgment of an acceptable level of risk. Thus, "safe food" involves a subjective evaluation of social issues and values, as well as a scientific assessment of risk.

Food Safety Regulation

Regulation of food safety was largely the responsibility of state and local officials until the first decade of the twentieth century. The emergence of a national market for food, combined with shocking stories of practices in the food industry, spurred the federal government to reassess its responsibility to ensure food safety.

Although those seminal laws provide the basic legal and institutional framework for the federal regulation of food safety that we observe today, they are very different in their implementation and responsibilities. Their core was similar: each prohibited the shipment in interstate commerce of food that met any of several definitions of "adulteration." Those definitions, or proscriptions, targeted products that were spoiled, contaminated with filth, derived from diseased animals, or contained unsafe substances. In short, both laws sought to prevent the sale of food that could offend consumers or cause them to become ill.

The original Meat Inspection Act and Pure Food and Drugs Act focused on different sectors of the food market and, from the beginning, supported different approaches to ensuring compliance. The Meat Inspection Act, which has been administered by the US Department of Agriculture (USDA) since its inception, requires continuous on-site factory inspections by government inspectors using sight, smell, and touch to detect problems. The USDA also administers the 1957 Poultry Products Inspection Act, which requires continuous inspection of poultry and poultry products and the 1970 Egg Product Inspection Act, which requires continuous inspection of the processing of liquid, frozen, and dried egg products. Animal carcasses and processed products cannot be shipped from the processing plant until they have been inspected and deemed to be unadulterated by the USDA inspector.

The 1906 Pure Food and Drugs Act was administered by the USDA Bureau of Chemistry, which was renamed the Food and Drug Administration (FDA) in 1930. In 1940, FDA was transferred to the Federal Security Agency, and in 1953 it became a separate entity in the Department of Health, Education, and Welfare, later retitled the Department of Health and Human Services. FDA, with a more varied and ever-expanding industry to regulate, has used a sampling strategy in which far fewer inspectors pay periodic visits to settings where food is produced, processed, or stored to verify compliance with its requirements or to uncover and punish violations. In 1958, when Congress enacted the Food Additives Amendment to the Food, Drug, and Cosmetic Act of 1938, the FDA was given premarket review and approval authority over chemical additives to foods.

The meat and poultry inspection system still requires sight, smell, and touch inspections, which are ineffective in addressing the current issues of food safety. Expert opinion that USDA inspection procedures do not adequately address the microbiological problems associated with meat and poultry led to the adoption of a new approach to inspection using the Hazard Analysis Critical Control Point (HACCP) system. In contrast to the reactive characteristics of traditional food safety strategies, the HACCP system focuses on preventing hazards that could cause foodborne illness by applying science-based controls at each step of the process from raw material to finished product. Following the recommendation of a National Academy of Sciences committee, FDA adopted the HACCP approach for seafood inspection in December 1995, and USDA began the implementation of HACCP for meat and poultry inspection in January 1997 with completion of the implementation by January 2000.

Agencies other than USDA and FDA-such as the National Marine Fisheries Service (NMFS) in the Department of Commerce, the Environmental Protection Agency (EPA), the Centres for Disease Control and Prevention (CDC), and the National Institutes of Health (NIH)-also have roles in food safety. NMFS conducts a voluntary seafood inspection and grading programme and does research on seafood safety. The EPA Office of Prevention, Pesticides, and Toxic Substances is responsible for risk assessment, product approvals, tolerance setting, and research on pesticide residues in or on human food and animal feed. State health departments in conjunction with CDC are responsible for the surveillance and investigation of illnesses related to foods, and NIH conducts research related to foodborne disease. Many state and local agencies are also involved in food safety regulation and inspection.

With so many agencies involved in food safety, decisions and priorities often focus on specific issues rather than strategies, and they are not always well-integrated. The system that has evolved in the federal government for regulating food safety is complex, fragmented, and cumbersome.

The Committee and its Charge

Congress directed USDA's Agricultural Research Service to contract with the National Academy of Sciences to conduct a study on the scientific and organizational needs of an effective food safety system. The Academy appointed an ad hoc Food Safety Oversight Commission composed of selected members of the Institute of Medicine's Food and Nutrition Board, the National Research Council's (NRC) Board on Agriculture, and the NRC Commission on Life Science's Board on Environmental Studies and Toxicology; in turn, the ad hoc commission convened the Committee to Ensure Safe Food from Production to Consumption.

The committee was charged with determining the scientific basis of an effective food safety system; assessing the effectiveness of the current US system to ensure safe food; identifying scientific and organizational needs and gaps in the current system at the federal level; and providing recommendations to move towards the scientific foundation and organizational structure of a more effective food safety system for present and future generations. If organizational changes were recommended, a subsequent study on implementing the recommendations was to follow, if requested. Given the constraints on time and funding, the committee focused on the role of the federal government in ensuring safe food.

The committee is aware that safe food is and always will be a moving target with respect to both changing hazards and changing degrees of acceptance of risk. This report is not and cannot be a comprehensive, all-inclusive discussion of these issues and adoption of its recommendations will not end the effort to make food safer. It is not possible to foresee how risk acceptance will change or how the problems will change in decades to come, but federal food safety efforts must be designed effectively to deal with what is known and what is not known.

EFFECTIVE FOOD SAFETY SYSTEM

The definition, historical significance, and current status of food safety in India as well as recent trends in and reasons for concern about foodborne hazards. This chapter turns to the features of an ideal food safety system and to a somewhat detailed description of its attributes. Constructing a model system establishes a benchmark that provides both a point of reference for judging actual systems and a goal to be achieved.

THE MISSION OF THE SYSTEM

There is universal agreement on the need for safe food, but no consensus on the means by which safe food is secured. The effectiveness of a food safety system begins with a clear, unified mission that focuses and integrates the varied needs and responsibilities of all stakeholders, gives the stakeholders a basis for achieving the goals of the system, and is broadly accepted.

The committee defines safe food as food that is wholesome, that does not exceed an acceptable level of risk associated with pathogenic organisms or chemical and physical hazards, and whose supply is the result of the combined activities of Congress, regulatory agencies, multiple industries, universities, private organizations, and consumers. The mission of a food safety system should be stated as an operational charge that uses and reflects that definition. The committee defines the mission as follows:

The mission of an effective food safety system is to protect and improve the public health by ensuring that foods meet science-based safety standards through the integrated activities of the public and private sectors.

The remainder of this chapter focuses on these matters. Activities that are implied in the definition and are incorporated in achieving the mission to ensure safe food include: adequate monitoring and surveillance; science-based research and development; incorporation of the tenets of risk analysis, including risk assessment, risk management, and risk communication; good practices in food production, processing, manufacturing, retail sale, transportation, preparation, and handling; and appropriate technical assistance and education. Box -1 provides a definition of science-based and related examples. A science base for ensuring safe food encompasses many elements. When utilized, these elements improve the ability to identify, reduce, and manage risks; minimize occurrence of foodborne hazards; gather and utilize information; enhance knowledge; and improve overall food safety. Several examples of science-based actions that have been implemented in the US food safety system that are readily recognized as positive elements of the system include:

- Implementation of low-acid canned food processing technology, which reduces the risk of botulism;
- implementation of HACCP systems and risk assessment in decision-making;
- approval of irradiation technology for use in spices, pork, beef, poultry, fruits and vegetables;
- prohibition of the use of lead-based paints on utensils that come in contact with food;
- estimation of maximum allowable exposure levels to pesticides;
- development of standards for allowable practices associated with transport of foods following transport of pesticides in the same containers;

- use of labeling as a device to warn consumers who are sensitive to potential food allergens of the content of the allergen; and
- requirements that meat and poultry products at the retail level carry consumer information related to safe food handling practices.

While the approaches above are important successful science-based tools in food production and processing, these are only examples of implementation of the scientific basis for food safety. An effective food safety system also integrates science and risk analysis at all levels of the system, including food safety research, information and technology transfer, and consumer education.

Ensuring a system to provide safe food within India is a common goal, and legislative mandates should direct the components of that system. The mission should be focused on using public health resources and management to perform risk analysis and research and at the same time on optimizing coordination and planning of prevention, intervention, control, response, and communication mechanisms.

General Attributes of the System

An effective food safety system is an interdependent system composed of government agencies at all levels, businesses and other private organizations, consumers, and supporting players. The system is dynamic and aligned to the unified mission of improving food safety so as to maintain and improve the public's health and well-being.

The interrelationships of an effective food safety system. Although the players have key, independent functions, they must implement many of their actions through strong partnerships. The system is built on the flexibility and adaptability of the players and on the nature and course of their relationships, and these affect their ability to prevent, identify, and resolve food safety issues in the most efficient, effective, and cost-beneficial manner.

The system is responsive to the expedient issues of today but also evolves strategically to meet future challenges. It must be effective in both the domestic environment and the new global food environment, with increasing international trade. The system is science-based with strong emphasis on risk analysis and the use of data. It is attentive to learning through the use of feedback loops and continuous improvement. Although responsiveness to and coordination of food safety crises are critical attributes, the system is designed to stress prevention and detection of emerging problems.

The system has adequate funding, is supported by strong research and education components, uses technology adequate to the task, and is integrated to achieve its mission. Statutory and regulatory authority promote the system's horizontal and vertical integration. Because integration is such a critical attribute, the system requires strong, centralized leadership. The system champions a culture of capacity building. With the transition towards a new scientific and risk-based foundation, agencies will encourage and fund retraining and further development of their employees, and they will initiate plans for the recruitment and retention of high-quality staff with the skills and knowledge to enhance the new system and its changing operations and focus. Capacity must also include consumer knowledge and practices, taking into account cultural sensitivities and practices.

An effective system is commensurate with today's driving forces, trends, and societal expectations. Partnerships will expand with the growing recognition that government cannot abandon our food safety problems to private industry and consumers.

The effective system stresses the inclusion of its players in their roles but also acknowledges the need for effective regulation, compliance, and enforcement. Although some of the roles will need to change, the system functions in an environment of trust and respect. The globalization of the food system and other factors that can increase the risk of foodborne incidents mandate urgency in adopting the system.

The effective food safety system is focused on public health, and its many actions are aligned to achieve a safer food supply, improve public health, and instill consumers' confidence in both the system and their role in improving it. Finally, the dynamic interconnectiveness also promotes the attainment by all of the players of both responsibility and accountability for making the food safety system perform optimally.

The Importance of Partnering

Food safety is the responsibility of numerous and diverse stakeholders, and partnerships provide the links that are necessary to build a coordinated and cohesive framework for action. Partnerships can improve efficiency and provide a mechanism for information and technology transfer. Interaction and communication through partnerships lead to cooperation and collaboration among public and private interests. Partnerships can also help to integrate regulated activities with important non-regulatory components of the system.

Incentives can greatly influence and facilitate the building of effective partnerships. "Positive" incentives are often financial. "Negative" incentives can include the desire to avoid legal or regulatory action or media attention. Another incentive for partnerships can be the generation of new and useful information that improves production and processing capabilities simultaneously with improved control of risk.

The market is an important incentive for private industry. As global markets and domestic consumers expect safer food, safety itself can become a factor in differentiating products. Retailers can help to leverage this concept as brand names become associated with reduced risks.

Some factors, such as intellectual property issues, pose challenges to the establishment and maintenance of partnerships. These issues might become increasingly difficult in the future, as such sciences as the microbiology of food continue to advance rapidly.Despite the challenges posed by the diversity of the players and changing priorities within the system, the potential realm of strong partnerships is large and includes all partners: government, the private sector, consumers, and support players. Partnerships should be formed and function in an open process that is independent and protected from political, economic, and social pressures. Partners must have clear delineations of responsibility and the authority to make decisions to meet responsibilities. They must have the resources to work together effectively. Successful partnerships are based on close, detailed, and accurate communication and collaboration.

In the public sector, the federal government is in the best position to influence how the other components of the food safety system work together. Its actions, which often take the form of regulation, originate in federal food law. The federal government must guide the system with national food law that is clear, rational, and based on scientifically determined risk. A few principles form the basis for ideal food law, which can be conceptualized by the philosophical structure of food safety. This structure includes a process that is fair and open to participation by all without political, economic, or social pressures and that provides for adequate authority and budgetary considerations. Legislation must be flexible and enforceable, and it must comprehensively address all aspects of the entire system from production to consumption. The authority and responsibility of the federal government and its interface with private and other partners must be well-articulated. Definition of a broad federal role is promoted by the similarities of food safety issues across states and demographic groups. Other levels of government have a role in shaping federal activities if they are to be effective in implementing national standards and in dealing separately with local issues.

With a sound food law in place, partnerships between the federal government and others in the system can establish a framework to provide several important functions. For instance, partnerships of federal, state, and local governments with industry, universities, private organizations, and consumers can ensure that the system is science-based and risk-focused, that surveillance and monitoring efforts provide sufficient information to maintain and improve effectiveness, that research and education efforts are properly focused, that regulation and enforcement are effective and consistent, that the system is responsive to new technologies and changing consumer needs, that a continuous process of evaluation can respond to the queries and problems of all stakeholders, and that resources are adequate and appropriately allocated throughout the system.

Risk Analysis

The scientific foundation for decision-making within a food safety system is risk analysis. The role of risk analysis in an effective food safety system is threefold: it provides a basis for identifying where resources should be allocated in the short term; it constitutes a mechanism for determining where public and private efforts should be directed in the long term, especially with respect to research and preventive measures; and it yields important information for estimating and analyzing the costs and benefits of policy alternatives.

The components of risk analysis-risk assessment, risk management, and risk communication-use interdependent approaches, methods, and models; none can function well in the absence of the others. Good risk communication is required for effective risk assessment and risk management. All components of risk analysis-scientific, economic, legal, behavioural, or other-require an environment that is independent of the derivation of policy and must be subject to agency leadership.

Focused risk assessment identifies risks that have the most important consequences for human health and (not always the same) shows where the most progress can be made with available resources. Scientific risk assessment should identify the risks associated with one or more possible actions, or risks associated with taking no action. Risk assessment is fairly well-developed for chemical hazards, such as pesticides, but not for foodborne pathogens.

Development of risk assessment for the complexity of such pathogens must be given high priority. A pivotal component of risk analysis is risk assessment. Scientific risk assessment is the process of determining the relation between exposure to a hazard at a specific magnitude and the likelihood of an adverse event or disease. The science should characterize the nature and magnitude of risks to human health associated with food related hazards.

On the basis of the scientific assessment, regulators can make informed decisions about where and how to allocate resources to prevent and control hazards. The same scientific technique can sometimes be used for a range of problems throughout the entire system. Uniformity in the process of risk measurement can reduce uncertainty about how specific foods or processes will be addressed, and thus benefit non-regulatory components of the system as well.

Adequate Surveillance and Monitoring

Strong surveillance and technical support provide an infrastructure to set priorities for research, education, and response. Food safety, perhaps more than any other health-related activity of government, demands a continuing and dynamic "moving picture" of the system, rather than periodic "snapshots." For successful leadership, the federal system requires the ability to identify and set priorities among potential food safety hazards and emerging issues. Surveillance and monitoring, education and research, and enforcement and regulation are all used to identify, set priorities among, and address concerns.

Surveillance is the first step towards building the capacity to detect and respond to sporadic instances and outbreaks of foodborne illness. Surveillance and monitoring are also essential for evaluating the system, identifying emerging issues and new trends, and assessing risk from the farm or sea to the table. Monitoring foodborne disease and related hazards can allow early and rapid detection of hazards and illnesses. Thorough investigation of outbreaks to determine their sources and causes can identify crucial weaknesses in the system.

Focused Education and Research

Part of the role of the federal government in ensuring safe food is to promote education and research. Continuous, high-quality research is needed to keep capabilities of regulators up-to-date and to ensure rational and maximally effective evaluation and decision-making as well as optimal preventive measures.

Effective education, the strongest and most important form of information transfer, requires partnerships at the federal, state, regional, and local levels. The federal government should lead efforts to educate and inform the general public on such issues as labeling (including health and nutrition information), standards and specifications, changing technologies, and advertising. All stakeholders in the food system should be fully informed of risks and of food-handling practices that reduce those risks.

Both the responsibilities and the needs for setting education and outreach priorities are shared among many stakeholders. Integration and collaboration can help to focus efforts where the needs are greatest. Priorities for education are closely tied to communication capabilities and should be based on need. General academic and land-grant institutions in partnership with private industry and a variety of organizations play critical roles in the transfer of information. Producers, processors, and consumers have needs for both general and specialized education.

Carefully planned education programmes to communicate fundamental issues are important, with additional opportunities for education in times of crisis. Both should be used during major or highly publicized foodborne illness events. Education about the critical role of consumers in the food safety system might be especially effective.

Research priorities should be focused on the prevention of foodborne illness; this requires research on measures of current problems to determine how the problems can best be brought under control. Specifically, research priorities should be related closely to needs as determined by effective surveillance.

Some research priorities in an effective food safety system must be responsive to emerging problems, but others must have a long-term perspective, and effective research management requires a capability to shift resources and emphasis. In the past, research priorities have often been based on opportunities or pressing issues of the moment. Typically, research funding has been event-driven, rather than evenly sustained in support of longer-term objectives. The press of current problems is an ever-present threat to the needed long-range research, and an effective food safety system will support and protect both. Research must always be relevant to societal concerns and to the complex problems of ensuring a safe food supply. Research priorities should be established internally and externally in partnership with private industry, academe, consumers, and other stakeholders. The priorities must be widely known and accepted; it can be difficult for any agency to maintain an appropriate long-term balance when the news media, the public, legislators, and others, direct issues to be set aside when a crises of the moment arises.

The research budget, especially for long-term projects, should be protected from rapid swings and decreases. Perhaps more than most other activities, the development of effective research teams requires development over time, with significant effort, patience, and resources. A single lean budget can destroy the benefits of effective team-building efforts over a period of many years.

Not every research need can be met, but a force of the best scientific research investigators are needed to solve the long-term, difficult, and highly complex issues of food safety. These experts must be dedicated to the mission of protecting public health by ensuring a safe food supply and they must not be distracted by needs for research in other fields, such as improved methods of food production or processing, or treatment of food related illnesses.

Consistent Regulation and Enforcement

Mechanisms for the prevention of foodborne illness can span the spectrum from voluntary to regulatory and from outright bans to warning labels and education. Regulation is often intended to be a preventive and protective measure. A prerequisite for any regulatory action should be the assurance of adequate, consistent, and effective enforcement. The ability to protect requires the authority and resources to take action. The food safety system must contain adequate provisions for enforcement of regulations and must clearly link responsibilities with accountabilities. Many challenges posed by the current food safety structure, result from an inability to take action, respond to needs, or protect. Consistency is important in all aspects of ensuring safe food. Similar risks require similar planning, actions, and response. For instance, the intensity, nature, and frequency of inspection should be consistent for foods associated with like risks.

Adaptation to New Technology

The food supply and food production, processing, distribution and consumption practices in the United States have changed dramatically during the past 70 years. Further major changes will certainly occur, though it cannot be predicted when or what they will be. The fundamental shifts from a few, largely local items to today's variety, from small neighbourhood stores to supermarkets, from familiar individual ingredients to prepared foods, from home to outside preparation or consumption, and from traditional to "ethnic" cuisine were not widely predicted before they were well in progress. As these shifts continue to occur, the technologies associated with production, processing, manufacturing, and delivery of food also continue to change.

Hazards associated with food consumption also are continually changing. In recent years, the risks associated with fresh fruits and vegetables have grown because of increased consumption, variety, and importation of produce from developing countries. Importation of different types and amounts of foods from outside the United States expands the federal responsibilities and expertise needed in the US food safety system.

The system must include primary prevention of new food hazards through research and other activities and be flexible to respond to new and potential hazards. The food revolutions of this century have proceeded irregularly and often simultaneously, but few have been adequately anticipated. It is certain that revolution will continue, and an effective food safety system will prepare for the known as well as for future hazards that cannot be clearly seen.

The changing attitudes, concerns, desires, needs, and roles of consumers dictate the characteristics of the food supply and its associated risks. The United States has become home to people of many origins and cultures. With those cultures has come the desire for particular foods. A great challenge to a food safety system is to meet the changing demands while securing safety of a rapidly expanding and diversifying food supply.

Human and Financial Resources

The immense responsibility of an effective food safety system is coupled with the need for adequate human and financial support. If federal and other programmes are to prevent, identify, track, control, and respond to food related illness and reduce risks of future outbreaks or hazards, they will require commensurate funding. Insufficient funding and lack of incentives can diminish an ability to inspect and control foodborne illness; curtail needed research; prevent sustainable, aggressive, and innovative education efforts; and create instability in the system. For instance, the role of states as a public partner in the food safety system has traditionally been important. State and local regulators can help to ensure the safety of food in commerce, and they can assume responsibility for monitoring and enforcing safe food-handling practices; but adequate state and federal funding is needed to carry out these tasks. The economic costs and benefits of reducing the risk of foodborne illness are difficult to determine accurately and consistently. Regulations that govern the production, processing, distribution, and marketing of food can create benefits by improving food safety, but they can also increase producers' costs and potentially raise food prices. The economic costs of inaction are difficult to assess because the data typically reflect only reported incidents and deaths; unreported events generally elude economic assessment. Net benefits of increasing food safety can be maximized by equating the marginal benefits of safer food with the marginal costs of achieving food safety goals. Market-oriented approaches to food safety include economic incentives. In addition, personnel must be trained, developed, and kept up-to-date with advances and innovations elsewhere. Scientific and managerial competence must be ensured to build the human capacity necessary to support the system.

Provision of adequate funding and staffing are not sufficient; the system must also be efficient and cost-effective. One cost-effective measure is early detection and early warning of outbreaks before they spread. Federal coordination with regard to research and education efforts can ensure rapid and efficient response.

Every system should be practical. The mission must be practical, the goals must be attainable, and the work must be feasible. Primary food hazard prevention approaches and research (both long-term and short-term) efforts must focus on practical applications. Inspection, surveillance and monitoring, and education programmes should be practical in their gathering and conveying of information.

The federal government plays a prominent role in ensuring that vital elements are in place and that the unifying mission of food safety is sufficiently addressed. The coordinated application of scientific, technical, and regulatory measures to prevent foodborne illness and promote public health and safety is critical to the federal government's mission, with its core responsibility to ensure an effective food safety system. There are also occasions where regulatory activity can encourage or require the use of new technologies, which causes movement towards higher standards in the system.

Private Sector Partners

Private partners working in an effective food safety system include producers and importers, processors, marketers (retail and wholesale), food services, trade organizations, professional societies, and private organizations.

The primary and common roles of the various private-sector partners include bringing food to the tables of consumers and bringing sound scientific information to others in the partnership. Within the food production system, the private sector has the primary responsibility for ensuring food safety. To be most effective in this task, the private sector must maintain close interaction with the public sector, which sets standards and provides oversight, and with the consuming public, which expresses its food needs and choices. The roles and contributions of specific private partners are varied and distinctive, and each can make unique contributions to the overall functioning of an effective system. Several examples of successful partnerships that highlight private roles in the framework of an effective food safety system are provided below.

Producers

Production in some areas may represent the least controlled point in the food supply system. However, there are valuable opportunities for effective partnerships between producers and governments in the food safety system. An example of a combination of regulatory and non-regulatory partnership between producers and federal and state governments is the National Poultry Improvement Plan (NPIP), in which state authorities cooperate with the US Department of Agriculture to administer regulations for the improvement of poultry, poultry products, and hatcheries.

In this example, industry participants follow both the general provisions of the plan and specific provisions that are related to their industry category hatcheries, dealers, and breeders. Food animal industries use and understand on-farm quality assurance or control programmes; these could be expanded to include food safety provisions if they are cost-effective and serve as an incentive to market products. The industries are to follow all sanitary provisions and undergo testing and inspection. The industries pay most costs of the plan and seek to achieve and maintain a free or "clean" status. The plan has been useful, especially in breeding flocks, and has helped to eliminate several poultry diseases and prevent the spread of disease from breeders to commercial flocks. More recently, the NPIP adapted an on-farm sanitation programme to prevent and control the serious human pathogen *Salmonella enteritidis* in hatching eggs and chicks.

The NPIP provides a working model that involves federal-state-industry partnership. Industry members use the findings to inform the public that they have met the highest standards and have free, or "clean," status, a designation that adds value to their products and is a "preharvest" animal-production model that might be used in an expanded version for other food animal species. Federal and state agencies benefit from reduced requirements for inspection and oversight of "clean" facilities.

Marketers and Distributors

The dairy processing industry recognized the need for effective control of foodborne pathogens by promoting the enactment of effective controls of milkborne disease, including pasteurization. To assist states and municipalities in their efforts to prevent milkborne disease, the US Public Health Service developed model regulations for voluntary adoption by state and local milk control agencies in 1924. A model milk regulation, the Grade A Pasteurized Milk Ordinance, was developed and implemented through partnerships between the dairy industry and federal, state, and local regulatory agencies.

Private industry can contribute to the shaping of the scientific basis of the future food safety system. For example, the Flavour and Extract Manufacturers Association and the Food and Drug Administration developed the FEMA Generally Recognized as Safe review programme to ensure the safety of food flavours and extracts. In another example of industry-FDA cooperation, the development of low-acid, canned food regulations was a major step towards the safety of canned foods by reducing risks of botulism. Those examples illustrate how a progressive, scientific partnership can contribute to a food safety system.

Education and information transfer can also be important contributions of private industry and its associated trade organizations. For instance, the Food Distributors International and its retail partners have engaged in a cooperative food safety initiative, implemented through the Food Marketing Institute. This project assists the allied food industries to share information, to keep up to date on regulatory developments and technical resources in food safety, and to develop and maintain procedures for the industry.

The egg industry has integrated government efforts, academic research, and other industries to address food safety concerns. The industry has implemented new technologies developed by government and academe to reduce foodborne illness, has developed model partnerships, and has educated and informed consumers about the safety and nutritional composition of egg products.

The Role of the Consumer

Consumers have a large and critical role in an effective food safety system. First, the food system revolves around the purchasing power and decisions of consumers. Annual expenditures for food are enormous—over $700 billion in the United States. More important, the health, well-being, and longevity of the nation's population ultimately depend on individual consumer decisions. As discussed earlier, it is the role of the private (industrial) sector to deliver to consumers food that is wholesome and safe and food that can be rendered or maintained safe by appropriate handling and preparation by the consumer. After food leaves the production and distribution systems, a different set of risks affects food that is prepared and consumed at home, for which only consumers can be responsible for safety, and food that is prepared by commercial establishments away from home (restaurants, takeout shops, grocery store delicatessens), for which vendors retain major responsibilities for safety, although consumers must still follow good safety practices.

As has been noted before, the food system and the food safety system have constantly changed in many ways to meet consumers' perceived needs. This evolution has depended on close interactions between government and the private sector, but the role of government is often unseen except during crisis situations. Continuing activities of the government and the private sector provide the foundation of a safe food supply, but they must also provide the information and education that consumers need if they are to meet their own responsibilities.

Although consumers' desires, needs, and cultures have changed, the basic standards of food hygiene in the home have changed very little and probably will not change much in the foreseeable future. The food safety system should continue to include fundamental information about frequent and effective washing, prompt refrigeration, and proper cooling. Consumer knowledge on good practices is essential to consumer action, including handling, storage, and preparation of food in ways that improve food safety. Consumers should have reason for confidence that food and food components are safe when received, and they should know how to protect their food as it moves through preparation and consumption.

The size and importance of the consumer interactions with the food system have contributed to the growth of consumer organizations. These play an important role, particularly in advocacy about food safety matters. In addition, they help greatly by providing education and information. They should be involved in the development of protocols and studies to fill the vital information needs of the food safety system. Their leadership role requires them to ensure that their positions on food safety issues are science-based, and that they fairly represent the concerns of consumers in their advocacy role.

BOX-2 The Consumer's Role in Food Safety

In the early 1900s, pasteurization of milk was becoming common. It was the consumer's responsibility to preserve the quality of milk after it was delivered to the home. The consumer was instructed that "as soon as possible after the milk is set on the doorstep, it should be taken in and put in a cold part of the ice box. If some time must elapse between delivery and care by the consumer, an insulated box should be provided to protect the milk from heat and/or freezing. Milk should be placed only in clean pitchers or receptacles. Milk that has stood at room temperature during a meal should not be poured back into cold, fresh milk, but kept in its own container." Keep clean and cold was the rule that was followed to preserve the quality and flavour of milk. Today, the rule of almost a century ago still applies to consumer responsibility, as is demonstrated by the Fight BAC campaign's educational message. New mechanisms for production, processing, and delivery have evolved, but consumers remain responsible for ensuring that milk and other products are kept clean and cold.

The Role of Other Partners

Government, the private sector, and consumers are not the only recognized organizations that are vitally involved in the food safety system.

Scientists, educators, and food specialists are also key partners who are placed throughout the system and in academic institutions. For example, the Institute of Food Technologists represents food scientists, and the American Veterinary Medical Association represents veterinarians in industry, academe, and government agencies. Both professional organizations have subgroups that interact with the government agencies.

The news media cannot be considered an active partner in a food safety system, but are important players. Objective, accurate reporting about food hazards and timely transfer of information to the general public are essential.

The news media also engage in efforts to educate the public about basic food hygiene. In an effective food safety system these groups are recognized for their distinct and important roles and are encouraged to participate by membership on advisory groups, by providing direct input to the regulatory process, or by other actions.

Interdependence

Ensuring safe food is an appropriate role for government, but that role is not exclusive, nor is it a contract between government and the American public to guarantee absolute safety. A strong government presence may be necessary for a strong food safety system but is not necessarily sufficient to create a stronger and better food safety system. An effective food safety system is based on a particular set of attributes and relationships and a focus on improving the public's health and well-being, and the roles of federal, state, and local governments must be well defined, coordinated, and consistent in achieving a single mission.

The public often turns its attention to what government *should* do rather than what government *can* do. But what can government at its various levels actually do to promote food safety, and what conditions must exist to optimize its performance? The answers to those questions are based on defining the functions of the food safety system that can best be performed by each of the actors and on the ability to fill in system gaps in light of the inherent limitations of public service today. What can others or other systems do just as well or better, and should they? An effective food safety system is built on fusing the strengths of multiple players, by linking roles and authorities to responsibilities, and by aligning shareholders' efforts towards a single overarching mission of public health. An effective system emphasizes partnerships new and old, integration, and the need for accountability of all participants.

The food safety system in the twenty-first century should be based on a framework of common goals, measures, rewards, teams, and networks that focus on outcomes. Once the framework is developed and operating to address desired outcomes, and once the attributes of a successful system have been integrated, optimum features of a final organizational design will be more apparent. There is a critical need to connect the food system strategically with an effective food safety component; this cannot be an afterthought but rather must be a planned and purposeful activity.

It is clear that the food safety system must function as a integrated enterprise. That is, it must be agile, fluid, connected, integrated, and transparent. Integration is the antithesis of a system with components in isolation, barriers, and disconnected functions, rules, and policies. One of the most important aspects of an effective food safety system is the ability to assume an active, preventive role, as opposed to a reactive mode, to deal with problems already present. Focusing on prevention of foodborne illness rather than on response to failures is key to identifying and controlling outbreaks of foodborne illness. The system must be built to resolve today's problems but also with a special contemplation and flexibility to address future needs that are not yet foreseeable in detail.

Most important, systems for continuous improvement must be based on information and science. Feedback processes are critical to learning, improvement, and change. The food system requires good data at numerous points to measure results and adapt processes.

A unified food safety mission implies that goals are adopted to protect the public health, reduce risks of foodborne illness, and maintain, build, or rebuild consumer confidence. Means for evaluation and feedback are needed to determine whether the food safety system is meeting its goals. The criteria for assessment of the system outcomes include verifiable measures of progress, such as reduced numbers of foodborne illness outbreaks, decreased incidence of chemical residues in food, and a better understanding of foodborne risks that improves public policy and promotes improved consumer practices.

Recommendations

Congress commissioned the National Academy of Sciences, through the US Department of Agriculture's Agricultural Research Service, to undertake the study that resulted in this report. The charge to the committee was to perform two main tasks: assess the current US food safety system and its effectiveness in addressing the continually changing concerns about food safety; and provide recommendations on scientific and organizational changes needed to ensure an effective food safety system for the present and future generations.

The effectiveness of efforts to ensure the safety of food for US consumers is not solely, perhaps not even primarily, the responsibility of the federal government. An effective food safety system depends on the collective efforts of food producers, processors, transporters, suppliers, preparers, retailers, and handlers; of officials at the local, state, and federal levels; and of consumers who select and prepare food. There are three overriding conclusions that the committee came to as a result of its deliberations:

1. An effective and efficient food safety system must be based on science.
2. To achieve a food safety system based on science, current statutes governing food safety regulation and management must be revised.
3. To implement a science-based system, reorganization of federal food safety efforts is required.

The committee's recommendations focus on two main areas. The first addresses the scientific basis of prudent, cost-effective food safety regulation. The second addresses the legal framework and management structure within which the scientific basis should be developed and deployed. The committee presents these recommendations as an integrated, mutually reinforcing package. Specifically, the committee believes that the development and, critically, the use of a scientifically supportable foundation for regulation is not likely to be achieved within the current scattered organization of federal food safety programmes, where responsibility is dispersed, budgets are separate, a unified mission is lacking, and no single official has formal or public overall responsibility and authority for performance.

More study is needed to determine the best organizational plan for a centrally unified framework for managing the federal system. The committee's recommendations do not assume that the only way to improve food safety is to create a single responsible agency. The committee also recognizes that designing the details of a centrally managed federal system will generate controversy and will in any case take time. Moreover, it should be emphasized that many of the specific elements of its recommendations can and should be implemented without waiting for the creation of a centrally managed federal system.

The committee emphasizes the urgency of reform, both scientific and organizational. Threats to the safety of food consumed by Americans are ubiquitous and, whether or not the threats are growing, they are certainly changing. Diverse observers of the food safety system, including entities attached to both Congress and the White House, have been calling attention to specific gaps and deficiencies for more than a decade. Responsive legislative and administrative actions along the lines of those recommended are long overdue.

Scientific Recommendations

The United States has enjoyed notable successes in improving food safety. An example of these is the joint government-industry development of low-acid canned food regulations, based on contingency microbiology and food engineering principles, that has almost eliminated botulism resulting from improperly processed commercial food. Similarly, the passage of the 1958 Food Additives Amendment to the Food, Drug, and Cosmetics Act of 1938 was a "technology forcing" event that improved the evaluation of the safety of added and natural substances and reduced the risks associated with the use of food additives. With increasing knowledge, many rational, science-based regulatory philosophies have been adopted, some of which rely on quantitative risk assessment. Adoption of this regulatory philosophy has been uneven and difficult to ensure given the fragmentation of food safety activities, and the differing missions of the various agencies responsible for specific components of food safety. This philosophy must be integrated into all aspects of the food safety system, from federal to state and local.

The greatest strides in ensuring food safety from production to consumption can be made through a scientific risk-based system that ensures that surveillance, regulatory, and research resources are allocated to maximize effectiveness. This will require identification of the greatest public health needs through surveillance and risk analysis. The state of knowledge and technology defines what is achievable through the application of current science. Public resources can have the greatest favorable effect on public health if they are allocated in accordance with the combined analysis of risk assessment and technical feasibility. It is important to recognize that limiting allocation of resources to *only* those areas where high priority hazards exist can create another problem: other hazards with somewhat lower priority but with a much greater probability of reduction or elimination will not be addressed due to limited resources. Thus, both the relative risks and benefits must be considered in allocating resources.

7

Rural and Sustainability Development in Agriculture

AGRICULTURAL DEVELOPMENT IN SUSTAINABILITY

Agricultural development has three distinct but related dimensions: the physical-technical, the economic-financial, and the institution-human. The physical-technical dimension addresses land utilization, agricultural technologies, research and extension, agricultural inputs, farm-to-market access, productivity and production maximization, and so on. Agricultural development from the economic-financial perspective is concerned with costs, factors of production, terms if trade, pricing policies, subsidies, incentives, credit, return on investment, market mechanisms, and the like.

The institutional-human dimension looks at knowledge and skills, organization and management, training, implementation capacity, social relations, politics, communication, motivation, participation, local government, public-private sector linkages, culture and values, historical experience, and so on. Agricultural development, as with much of the development field, has tended to be the province of the technical and economic specialists. Their language and concerns have dominated the analysis of rural and agricultural sector issues, the specification of agricultural development objectives, and the design of programmes and projects to achieve those objectives. The institutional-human dimension has frequently been either completely ignored or treated as a source of problems or constraints to achieving technical and economic targets, a residual category for anything not defined as technical or economic.

For example, a World Bank staff member notes that, "between 1976 and 1980 the Bank invested U.S. $920 million in agricultural projects without funds in any one of them being allocated to the staff training needed for the new activities on which those projects depended".

Recent work by Ruttan represents one effort to build theoretical integration of the three dimensions. On the applied front, the USAID-supported farming systems approach to agricultural development is one example of an explicit attempt to integrate the three dimensions operationally.

Recognition of the importance of institutional and human factors in socio-economic development in general has increased substantially over the past 15 years or so, and is now a high priority area of attention by donor agencies and developing country governments alike. Being a social technology, institutional development does suffer in the eyes of crop scientists, agricultural economists, and macroeconomic analysts from a lack of specificity, predictability, and hard-and fast rules.

Despite its "softness," institutional development is a frequently espoused need of developing country officials and an aim often expressed by donor agencies. U.S. foreign assistance policy, for example, aims officially "to build and maintain the social and economic institutions necessary to achieving self-sustaining growth".

This goal of institutional development is nowhere more important than in the rural and agricultural sectors, which continue to be the source of most employment in developing countries. Except for emergency relief, where the delivery of food for consumption is unavoidable, donor policies stress that assistance for the countryside should be an investment, that it generate a flow of benefits that endures after the external funds run out.

The cumulative lessons of experience show that to have lasting impact, donor efforts to improve rural areas in the developing countries, and agriculture in particular, need to work with durable indigenous institutions both at the central and local levels. It is not surprising that, in the 1990s, institutional sustainability is emerging as one of the major problem areas for sectoral management, as well as for public administration of core government functions in developing nations.

This chapter introduces the topic of institutional sustainability, focusing in particular on the agricultural and rural sectors, though its observations apply equally to other parts of the economy. To set the stage for the rest of the book, we start by briefly exploring the meanings of the key terms "sustainability" and "institution." Then we review the latest quantitative studies of sustainability, putting them in a comparative perspective.

The broad topic of development "sustainability" is obviously not a new issue. The folklore of foreign aid is replete with tales of equipment that never worked in the field and was left to rust, of well-intended recommendations that local people disregarded, of organizations that succumbed to apathy — if they ever functioned in the first place.

Critics of development assistance have always complained about this sort of waste. It has often been easier to be clear about the failed and was not sustained than to search out what worked, what lasted, and why. In the mid- to late 1980s, however, sustainability has surfaced as a distinct and pressing concern of policy-makers in donor agencies and in the developing world.

As the last decade of the twentieth century gets underway, the expansion of cropped area and the adoption of modern farming techniques are putting new, perhaps irreversible, stress on the natural environment in developing countries, while soft international markets for agricultural commodities are forcing reappraisal of the possibilities for the Third World to increase its farm exports. Conventional development strategies seem more and more to have run their course, yet fresh approaches have been difficult to find.

"Aid weariness" among the donor countries and high levels of indebtedness among the recipients have constrained the search for creative solutions. Finding ways to yes both external and indigenous development resources more efficiently to achieve broader and more lasting impact has become, accordingly, more difficult and more imperative.

Each of the academic disciplines cuts into sustainability problems from a different angle. Three points of view stand out, each reflective of the three dimensions of agricultural development previously noted. Agricultural and natural scientists tend to frame the problem of sustainability in terms of the long-term impact of current farming practices on the global resource base.

The chemicals and intensive cultivation associated with green revolution technology can degrade the environment and waste energy resources; scientific breeding of seed may reduce the native genetic diversity of crops, making food supplies more vulnerable to plant disease and pests. Because of such problems, agriculturalists and naturalistst are increasingly concerned with identifying and promoting methods of cultivation that can protect and enhance nature's assets, while continuing to produce enough food to satisfy burgeoning populations.

Thus the blue-ribbon Bruntland Commission has defined sustainable development as "development that meets the needs of future generations without compromising the ability of future generations to meet their own needs". Countries such as Nepal and Haiti are living (or dying) examples of unsustainable development, and represent cautionary stories for those who would ignore the ecological costs of detrimental policies and practices.

A downward spiral of events can result, with deforestation and erosion leading to food insecurity, extremes of poverty, class warfare, and ultimately to societal breakdown. Sub-Saharan African nations also appear to be in danger of proceeding down this road.

Economists, by contrast, are somewhat less preoccupied with the ecological dimensions of sustainability per se, since the mining of natural resources can be justified from an economic point of view. Their concern is not conservation by itself, but the appropriate tradeoffs between economic growth and environmental preservation.

Accordingly, economists tend to be troubled by the manmade market distortions that curtail the growth of production and employment in the Third

World. The most widespread examples of these distortions are the ill-considered of politically convenient price regimens that penalize the rural sector in many developing countries. There are extended economic costs to such policies.

As the World Bank notes: "Discrimination against agriculture on a sustained basis not only reallocates resources within agriculture but also draws them out of it. As labor and capital move out and technical progress slow, the longterm losses can be large."

Management experts tend to make the development project or the organization their unit of analysis, and to worry about sustainability in terms of how to meet recurrent costs or permanently improve capacity for implementation. Development project management has emphasized achieving planned targets within a specified time frame and budget, thus responding more to donor agency needs for accountability than to recipient country capacity-building requirements.

Donor-supported projects, with their autonomous implementation units, frequently experience a collapsing "balloon effect" when external funds and advisers are withdrawn. Because they are often organizationally "quarantined" and protected, with special budget and staff, such projects may leave little lasting imprint on regular government agencies, and in many cases may actually weaken them by drawing off the best national talent.

The management perspective on sustainability, however, has evolved away from a narrow project focus, expanding to concentrate more on programmes integrated into developing country agencies and on the question of how to maintain an ongoing flow of goods and services to intended beneficiaries. From the administrative point of view, sustainability means building organizational capacity to perform over an extended time period (longer than the life of-project horizon) and assuring the continuation of useful and valued outcomes for client groups.

Specialists concerned with nonagricultural sectors cut into sustainability issues in their own way. Health experts, for example, are usually more concerned with the clientele than the implementing agency; they see the problem in terms of promoting affordable preventative health measures, a opposed to curative care alone, measures that enable poor people to enhance their physical well-being over the long run.

But the fundamental problems of sustainability in health and other fields are closely parallel to those facing farmers and rural administrators. The thread that connects the agro-environmental, economic, and management perspective on sustainability is effective social institutions, which much be in place to deal with all the current challenges to development.

Overcoming the environmental damage of modern farming, for instance, is likely to require (among other things) that scientific organizations attack novel research problems, that private enterprise produce greater quantities of more sophisticated farm inputs, that farm advisory services disperse new knowledge, that local authorities mobilize farmers to change collective and individual behaviour, and that farmers' organizations collect and market unfamiliar crops.

Even something as seemingly straightforward as price as price reform requires considerable institutional capacity to analyze and monitor the impact of new prices, let alone to manage any political backlash. The institutional dimension, for example, has emerged as critical to the success of the structural adjustment programmes and sector adjustment loans provided by the International Monetary Fund and the World Bank.

In virtually all problems of "development sustainability" the institutional dimension plays a key role. Unfortunately, many Third World countries lack the depth of institutions to carry through the complex and exacting duties likely to be needed to tackle the development tasks of the 1990s.

RULE-ORIENTED INSTITUTION

What are institutions? The concept is subtle and thus subject to confusion. According to Huntington definition, they are stable, valued, recurring patterns of behaviour. Institutions thus include rules or procedures that shaped how people act, and roles or organizations that have attained special status or legitimacy. Their importance is now being rediscovered by the social sciences, including the development subfield.

An example of a rule-oriented institution is a system of land tenure; whereas a role oriented institution could be the legal authority established to adjudicate disputes arising out of the land-tenure system. Both rules and roles can be institutionalized, the former as codes of law or custom, the latter as concrete organizations.

It is useful to keep in mind the distinction between these two types of institutions. Development assistance for agriculture and rural development is sometimes aimed at altering a village society's fundamental rules, for instance, by promoting tenure reform and land redistribution. When practitioners seek to build better institutions, however, the role-oriented, organizational definition is usually what they have in mind.

The USAID-sponsored institution-building literature of the 1960s, for instance, converged around organizations makes this matter of official policy, arguing that "because institutions become tangible only through the policies and actions of particular organizations, much of [our] institutional development effort will be focused on improving the policies and procedures of key organizations."

Most other donors and developing countries adhere to a similar definition, often narrowing it to mean formal collectives in the public sector, such as ministries, parastatals, agencies, or commissions. This book follows these conventions part way, by focusing on institutions in the organizational sense;

however, our framework of analysis allows, indeed requires us, to simultaneously consider the effect on organizations of established codes of behaviour.

The development community's focus on role-oriented, organized activities unfortunately introduces another point of confusion, since "institutions" in this sense can be used loosely to refer to any formal or semi-formal collective entity. But not all organizations are institutions, any more than all institutions are organizations. As Perrow points out: "Some organizations are merely... rational tools in which there is little personal investment and which can be cast aside without regret.

Others become institutionalized. They take on a distinctive character; they become prized in and of themselves, not merely for the goods or services they grind out. People build their lives around them, identify with them, become dependent on them." Many rural organizations fail this test. When one refers to them as institutions it is often to speak hopefully about what they might become, and not realistically about what they, in fact, are.

A prime example of this failure is the experience of many developing countries with agricultural cooperatives, where the gap between what the coops were intended to be, and what they actually are, has been wide.

Our contributors concentrate on formally-constituted collectivities, but without prejudging whether they deserve the appellation "institution" in a strict sense. Further, the fact that the case studies tend to neglect informal modes of social organization, should not be interpreted as indicating that such institutions are less important in development. Nor should the focus of the case studies on discrete institutions be taken to mean that one may ignore the contextual factors within which organizations functions. To understand sustainability, it is important to keep this larger environment clearly in view.

INSTITUTIONAL SUSTAINABILITY

What are *sustainable* institutions? Strictly speaking, the term is redundant, since institutions are by their very definition, sustained patterns of social organization. But as a practical matter, development administrators have in mind collective entities that meet one or more of the following criteria: (1) they are able to recover some of their costs or even become self-financing; (2) they supply a continuing stream of benefits; and (3) they survive over time as identifiable units.

Whether these are sufficient or even necessary criteria of sustainability, however, is often left unclear. Certainly, organizational longevity by itself is a dubious measure, since it doesn't answer the question of how long a period is required before an organization can be considered sustained. One relevant consideration is whether it has passed through the crisis of succession, if it has accomplish what Weber calls the "routinization of charisma," so that it can perpetuate itself no matter who holds its offices. Unfortunately, knowing when leadership has taken root is not always possible *ex ante*. Self-financing is also a questionable indicator of sustainability, since developing countries are home to many organizations that provide essential services, but whose clientele are so poorly endowed that they cannot perpetuate themselves without indefinite subvention from overseas.

Desirable development entities may not be sustainable in this narrow sense. The question of long-term benefit flows is similarly unclear. It sidesteps the critical issue of who the beneficiaries of an organization are, for example, local landowners versus landless laborers. Elite-oriented collectivities are, almost inevitably, more likely to supply their members with a stream of advantages.

Given these difficulties, we would recast the phenomenon of sustainability in systems terms. We propose the following simple and widely applicable definition: Sustainability is the ability of a system to produce outputs that are sufficiently well-valued so that enough inputs are provided to continue production. This expression of the term adequately sums up what it takes for an institution to sustain itself.

Readers familiar with the social science literature will recognize the counterpart of the notion of "institutionalization," which Huntington defines as "the process by which organizations and procedures acquire value and stability." The important points to keep in mind are that institutions, properly understood, always serve one or more client groups, that to prevail they need to "keep close to the customer," to use private sector terminology, and that they become vulnerable when they fail to produce goods or services economically.

We want to stress that we imply no value judgment in our definition of sustainability. The concept, as we use it here, is analytic, not normative. Sustainability only requires that someone value an institution's outputs, so it may or may not be desirable for development. Our intention here is to explain why institutions sometimes persist and sometimes fail, not to ascertain whether they ought to do so.

PAST RECORD OF SUSTAINABILITY

Most observers would agree that too few of the rural development organizations supported by international donors over the past several decades have sustained themselves and become institutionalized. But can a more precise figure be put on the problem? USAID recently had 212 project evaluations reviewed from the perspective of sustainability.

Twenty-six percent of the projects earned strongly negative ratings, 56 percent got marginal marks, and only 11 percent of the projects were considered to have strong prospects for being sustained after the termination of U.S. assistance. Agricultural and rural development projects fell disproportionately into the least sustainable categories. The World Bank did its own study of this issue, using a different methodology but reaching almost the same conclusions. The bank reexamined the impact evaluations done on 31 projects during 1979-1985 (27 of which were in the agricultural sector) to determine how successfull the projects were in keeping up their activities.

Occurring about five years after a project is ended, an impact evaluation is a better gauge of sustainability than a normal performance audit. The study concluded that a majority of the projects were either unsustainable (32 percent) or marginally sustainable(26 percent).Only 42 percent of the projects had successfully achieved sustainability.

These two studies of institutional sustainability help quantify the problem, but their findings make little sense without a point of comparison. The fact is that organizations always have high attrition rates, even in a developed country like the United States. No sector is immune. Consider, for example, the sustainability of commercial enterprises.

A profit making organization that provides insufficiently valued outputs usually goes out of business, and this happens with great frequency in the United States. The consensus among academics is that 65 percent of start-ups fail within the first five years. Even the survivors the difficulty making the transition out of an entrepreneurial mode where their founders monopolize most important management functions. Few business (the *Fortune* 500 perhaps) ever attain the status and permanence that characterize institutions.

New not-for-profit ventures also face long odds. Consider the case of American colleges: of the 516 institutions of higher education founded in the United States before the Civil War, 81percent had ceased to exist by the 1920s. The same is true of voluntary associations, which are notoriously ephemeral, ebbing and flowing with popular preferences and needs. The history of the Grange, America's most notable farmers' organization of the nineteenth century, is illustrative.

Following the first lodge's establishment in Washington, D.C. in 1868, more than, 24,000 lodges formed over the next seven years. The peak year was 1875, but already some 5,000 lodges had lapsed into inactivity. Grange membership, which topped out at 450,000 during this period, shrank to 65,000 by 1880.

The sustainability of public sector is more difficult to determine, for poor service and dwindling clientele can be offset and masked by subsidies from general government revenues. Indeed, one of the leading gripes conservatives have about government is that its agencies and offices are sustained artificially. Being insulated from market competition, public bureaucracies can tolerate considerable inefficiency and lack of innovation, yet still survive.

Nonetheless, they are not immortal. Kaufman has studied this equation in the U.S. federal government. Of 175 administrative organizations extant in 1923, 15 percent had been terminated by 1973. To summarize, American organizations — whether they be formal or informal, private or public, profit-oriented or charitable — have had to fight uphill to get themselves institutionalized.

Obviously, none of these examples from the United States can be likened exactly to newly established or reconstituted organizations in developing countries today. This advanced country's experience, however, does give a rough idea about institution-building prospects in the Third World. Recent accomplishments there are probably more noteworthy than generally recognized.

The sustainability of organizations, it is important to note, is not an end in itself, but a means of achieving other developments goals. Some organizations ought to disappear, either because they were set up to do a time-bound task and are not needed on a permanent basis, or because they prove unable to perform useful and valued functions. Since development projects are often policy experiments institutional failure can serve a learning purpose, too, though national and international development agencies do seem to have particular difficulty avoiding past mistakes.

That unpredictable ventures such as development projects frequently prove unsustainable is inevitable. None of these observations should be grounds for complacence about today's institutional development strategies, but they do counsel for humility about whatever approaches are tried tomorrow. There are no organizational nostrums.

Institutional sustainability is not always a feasible objective, nor even necessarily a desirable one in some instances. Insurmountable external and internal problems can terminate almost any formal group. Yet, there is a clear need to improve the odds for sustainability in many overseas development activities. And such improvement is possible, at least in the broad middle range of country and project settings. Enormours experience has been accumulated over the past forty years about what sorts of interventions do and do not work in the long run.

Yet, many of the same mistakes seem to get made over and over. Organizational memories are short among the foreign aid agencies, and bias exists in many quarters for institution building of the "bricks and mortar" variety, irrespective of the capacity to actually run such institutions.

For developing countries and international donors to increase their success rate, they need to rethink their approaches to selecting, strengthening, and supporting institutions that serve development purposes, whether in the public, private, or voluntary sectors.

This task involves looking anew at project and programme design and level of integration into existing organizations; at the links between management training and performance; at decision making, leadership, and managerial choice; at developing institutional development strategies at the interorganizational level; at evaluating environmental conditions realistically, anticipating change, and not overreaching institutional limits, and so on.

SUSTAINABLE DEVELOPMENT POLICY

An example of a holistic and sustainable development policy based on human ecology can be demonstrated by the situation on the Andaman and Nicobar Islands. The suggestions developed below represent a step in the direction of policy-orientated counselling; as such they target application and implementation.

DATABASE

Information on the Andaman and Nicobar Islands has been sketchy and obscure, and the database limited. The first information recorded on these islands came from Claudius Ptolemaeus, the great Roman geographer, who lived in Alexandria in the second century. Ptolemaeus, who had never seen Andaman or Nicobar, described the islands as "Islands of the Cannibals".

The next reference to these islands is found in the writings of I'Tsing, a Chinese traveller who arrived in India in AD 673. He referred to the islands as "Andaban" and also described their inhabitants, evidently on the basis of hearsay, as cannibals. Marco Polo, the famous Venetian traveller, journeyed near the islands around 1260. He described the inhabitants like this: "All the men of this island have heads like dogs, and teeth and eyes like dogs; for I assure you that the whole aspect of their faces is that of big mastiffs".

Despite these obvious historical inaccuracies, there is nevertheless abundant written information on these islands. There is also much expertise to be drawn upon in Port Blair itself, and there is in fact a small, but accurate, firsthand empirical database to supplement the previous ones, that is observational impressions.

Port Blair was founded by the British as a penal colony for their Indian colonial empire. During the Second World War, the islands were occupied by the Japanese. They passed to independent India in 1947 as a centrally administered union territory with limited sovereignty.

The Andaman and Nicobar Islands are located 1200km east of the Indian subcontinent and are the remains of a former land bridge that extended from Burma to Sumatra. They extend between 6° and 13° on the northern parallel, thus falling in the tropical rain zone with 200 days of precipitation per annum. Dense tropical rain forests are the dominant form of vegetation on these almost unspoiled islands. Along the coast, they are encircled by mangrove forests. Coral banks in shallow sea stretch out in front of white, sandy beaches. The entire area encompasses 8300km^2.

Most of the approximately 550 islands are uninhabited. The Indian census estimates the number of inhabitants in 1981 at 190000, up from 50000 in 1971. Of these, 22000 are negroid descendants of the island's original inhabitants, and Asian immigrants of a tribal character, some of whom continue to live in complete isolation from civilization.

The rest of the population consists of the so-called locals, descendants of the Indian prisoners in and around Port Blair, as well as Indian immigrants and administrators from the mainland who have formed the political and economic elite since 1947.

Twenty per cent of the population is engaged in agricultural production. The chief crops grown on the 48 000 ha of cultivated land are rice and vegetables, followed by coconut and fruit trees. Tropical wood is processed in saw mills: in 1984 this amounted to 182000m^3. Despite the existence of a plenitude of sea life, the fishing industry has hardly developed. Except for the saw mills, there is no industrial production.

Some Political Proposals and their Supposed Social and Ecological Impacts

Before I visited Port Blair, the capital of the Andaman and Nicobar Islands, in 1987, I tried to sort out those political proposals that could have major impacts on the future development of the islands. The basis of most of my information consisted of rumours, newspaper articles, and information gathered from consultations with experts.

According to these sources, three major development proposals could be identified:

- A proposal for a defence area to test missiles,
- A proposal for a tourist centre,
- A proposal for a free-trade zone including a duty-free port.

Missile Testing Site

According to *India Today*, the Defence Ministry had been trying for the past four years to set up a National Test Range (NTR) in Orissa, Balasore District (estimated cost: Rs30000 million). Because of extremely strong opposition from local inhabitants it was suggested that the NTR be located in the Andaman Islands where fewer individuals would be affected — and possibly where opposition would not be so strong.

There are defence and budgetary arguments both for and against the Andaman Islands as a testing site, but more relevant to my purposes is a consideration of why the Orissa population resisted so strongly, because the reasons may prove valid for the Andamans as well.

For the Orissa farmers, "the area, spread over 400km^2 along the seashore and comprising 132 villages, is a virtual gold-mine with cash crops like betel-vines, coconuts and cashew besides profitable fishing in the shallow waters...". Local farmers claimed that the NTR project would destroy vast paddy fields, 30000 betel-vines, thousands of coconut palms and cashew plants at a value of over Rs1750 million.

Thus, the planned resettlement of farmers and their families could not only bring about the destruction of an ecosystem, it could also destroy a human ecological system where societal groups live in harmony with nature. This same argument could, of course, be applied to the Andaman or the Nicobar Islands should the NTR be located there.

Tourist Centre

What is tourism? A tourist was previously understood to be a person who was curious about the nature and lifestyles of other persons in other parts of the world. It was this curiosity that made the tourist travel. Because of this simple definition, we sometimes still believe that tourism promotes understanding between peoples.

Today, however, the tourist has been degraded into a foreign-exchange-carrying animal from whom local inhabitants or governments attempt to extract as much money as possible. Because tourists are willing to surrender to this exploitation, they are permitted to retaliate by exploiting the local environment and disrupting local social structures.

For example:

- Trekking in Nepal has had severe environmental consequences that threaten traditional Nepalese society.
- Coastal tourism in East Africa led to rising expectations among the local population. Children began to earn much more than their fathers as hard-working fishermen had ever earned; social hierarchies were upturned.
- Sex tourism in Thailand has disrupted local cultural values and beliefs.

What do these examples show? Again, although the impacts may be more subtle than those of the proposed NTR, human ecological systems will be affected. The impacts may be for the better or the worse, so the question should be raised as to how these systems can be affected for the better, that is, how can the greatest damage be avoided?

Ghetto tourism, where tourists living in a camp are kept apart from the indigenous population outside, was a partial answer to this, but ghetto tourism seems to be on the decline. According to Willibald Pahr, Secretary General of the World Tourism Organisation (WTO), experiences with ghetto tourism have not been favourable.

Another answer to this problem was given by countries like China, Burma and Tanzania. To cite one example, Gertrude Mongella, Minister of Tourism in Tanzania, stated that her country was opposed to mass tourism and, consequently, sex tourism does not play a role. Tanzania's tourist policy is to make the country attractive for educated persons who show a specific interest for the local countryside and local culture.

Interestingly, the Tanzanian tourist policy led back to the original definition of a tourist, which was thought to be obsolete — a person who is curious about other people. It is also worth noting that this concept of tourism had an elitist element to it. Moreover, this elitist way of thinking was being propagated by socialist countries — an amazing switch of values! My question is, therefore: is elitist behaviour necessary to protect human ecological systems, that is, to protect social systems and their natural environments?

Free-trade Zone and Port

For years, there has been talk of making Great Nicobar a duty-free port. Hong Kong's future remains uncertain, Singapore is crowded, and Great Nicobar is situated on the trade routes between India, Burma, Southeast Asia and the Far East. What role does Hong Kong play in this? It could either serve as a model for Great Nicobar or it could provide the needed capital for development as a result of its own political and economic uncertainty.

It is not likely that Hong Kong will provide the source of capital for Great Nicobar. First, those Chinese who were able and willing to leave the crown colony have already done so and made new homesteads in the Caribbean. Secondly, Hong Kong is flowering economically at present.

Thirdly, and most importantly, there is no ethnic Chinese minority on the Andaman or the Nicobar Islands. There is a strong Hindustani-speaking tradition on these islands, so the chances are better that they will attract floating Indian capital, including that of Hong Kong Indians who will lose their national identity in 1997 when the crown colony becomes part of the People's Republic of China. There are many possible objections to the proposed free-trade zone, but one is particularly significant. On the Nicobar Islands, there is still a growing aboriginal population. These islands are one of the few areas in the world that are almost untouched, with a functioning human ecological system virtually free from outside influences.

For this reason, the Indian Anthropological Survey protested against the institution of a free-trade port. Whether, as Iqbal Singh suggests, it would be possible to make Great Nicobar a duty-free port but leave the other islands untouched, is still an open question. It is not my opinion that all efforts to develop the Andaman and Nicobar Islands are a *priori* wrong. However, I do believe that those responsible must be extremely careful. There is no government in the world that can easily afford to destroy human ecological systems. Any recommendation should take this into account.

The costs and benefits of development must also be calculated, but not in purely monetary terms. What would be the price, for example, of the Nicobarese human ecological systems, which, once they are destroyed, cannot be rehabilitated. Finally, it will be necessary to differentiate between who will share the benefits and who will bear the costs of development.

Three Models for Future Development

The following recommendations represent an attempt to present some ideas — following the concept of holistic human ecology — concerning the future prospects of the islands, albeit based on limited observational data. This includes possible benefits to their inhabitants. I shall therefore proceed with a good deal of caution.

As a result of the information collected, three basic models of future development would seem to apply to the Andaman and the Nicobar Islands:

- An ecological preservation/social stagnation model
- A macropolis model, and
- An ecodevelopment model.

Ecological Preservation and Social Stagnation

These islands represent one of the very few, perhaps the only, specimens of truly natural ecosystems that have hardly been influenced anthropogenically. It is argued, therefore, that these islands must be preserved as they are. This has consequences for the three population groups on the islands, the tribals, including an aboriginal population, the so-called "locals" and recent immigrants, including government and administrative officials.

Some islands are populated with aborigines who live in harmony with their island environment. In view of the present slaughtering of Amazonian peoples, the aboriginal inhabitants of the Andamans and Nicobars may soon become the last undisturbed autochthonous communities on Earth and, as such, their lives, lifestyles, culture and environment ought to be preserved.

To achieve this, outside influence should be kept to a minimum, but this can only be realized if "civilized" humans are not permitted to disturb the aborigines. The "locals" should be confined to Port Blair and a few "civilized" peripheral spots. This implies that no further colonization of the islands will be permitted and that immigration is ruled out.

It further implies social stagnation in order to preserve an original human culture. It is obvious that this model cannot tolerate development that would include military ventures, capital- and technology-intensive endeavours like a free-trade zone, or any significant form of tourism.

Macropolis

This model would make a new Hong Kong of Great Nicobar. A possible five million people living on some 860km^2 would result in a population density of 5800 per km^2. In view of this figure, it would not be too far-fetched to say that more islands would be needed, be they for reasons of pleasure or for agricultural supply; such a development creates its own dynamics.

According to the proponents of this model, the development would be desirable because of its expected pull effect on the mainland economy. The anticipated economic benefits would outweigh the social costs of development that — in "extrapolating" the present scenario — could include the following: the "locals" will lose their cultural identity; the Shompen, the tribals on Great Nicobar, will disappear; the Nicobarese will decay and their numbers will dwindle. The economic benefits, it is argued, would also outweigh the ecological destruction suffered by Great Nicobar and the other islands. To the contrary, it is argued that human welfare is more important than nature protection, and that the economic and social benefits of the macropolis model will reach the locals and Nicobarese through the trickle-down mechanism, once these peoples have been "civilized".

Ecodevelopment

The ecodevelopment model is located between the extremes. Ecodevelopment has two roots: ecology and development. Therefore the contradiction between development or growth and environment becomes the dialectics of development plus environment. It implies that accelerated economic development be, at the same time, ecologically and socially sound or at least benign.

Ecodevelopment policy begins by considering the needs of the people. Thus, the elimination of poverty is a precondition of economic development, not its result. Similarly, the priority no longer lies in export earnings, but in creating employment in the domestic economy and in enhancing a broad income distribution. Employment includes selfemployment at the monetary as well as at the subsistence level.

The planning of ecodevelopment projects is shaped by several factors. First, the sociocultural values and behavioural patterns of the peasant family, and the politico-administrative scope of the village community are major determinants. Secondly, the economic limitations, in terms of low income earners who may be partially embedded in a subsistence economy, are also important.

Finally, the biophysical environment may impose natural constraints. Selecting an appropriate location for a project is governed by the ability of those responsible for the planning process to harmonize sociocultural, economic and ecological factors, which may require democratic resolution of social conflicts.

How does this concept translate into action for the Andaman and the Nicobar Islands? Let us consider the answer to this query in terms of the three elements, needs, self-reliance and environmental compatibility.

THE APPLICATION OF AIM TO THE INSTITUTIONAL SUSTAINABILITY INITIATIVE

AIM emerged as a product of a USAID-funded applied research effort, the Institutional Sustainability Initiative (ISI).

The Action Stream of ISI

During the 1980s, the ISI emerged from the recognition that sustaining the benefits of donor financed development efforts was a major policy problem.

At USAID's 1987 conference on the future of agricultural and rural development in Asia and the Near East, sectoral specialists from the agency's field offices (missions) emphasized institutional sustainability as a priority. In response USAID pledged to develop practical field guidance on this complex policy issue. Thus, the preparatory phase of the action-inquiry stream for the sustainability policy issue began; with a field level call for action and a USAID-Washington response. USAID's Bureau for Asia and the Near East took a leadership role in the ISI, and provided some initial funding. The ISI stressed "high quality and practical field guidance"; minimal funds were available for research. What research would be done had to be low cost and closely tied to practical results.

The next step in the preparatory phase was to obtain field level commitment to the ISI. Field missions were encouraged to request a sustainability assessment and design for their new and ongoing projects in the agriculture and rural development sectors; few projects had been designed with serious attention to sustainability.

In the investigation phase of AIM, the conceptual model devised in step one is systematically applied in the field to identify sustainability and conclusions. Informal contacts with field missions facilitated the selection process, and invitations for applying the model were plentiful.

The planning for the initial ISI applications was somewhat ragged. The first field teams went out while AIM was still being developed, and their instructions were minimal. Experienced consultants and USAID personnel were left to their own devices more than hindsight indicates was optimal. Some misunderstandings occurred. Changes were instituted later to assure that field teams were better oriented prior to their departure for the field. The SCOPE framework was used during orientation in two recent field cases, Thailand and CARDI, to orient the ISI team.

For each field site, a scope of work for the sustainability consultancy was negotiated, which melded the goals of the donor, the mission, and the assessment/design team. The team met prior to departure, and planned their activities, adapting issues and questions to the field setting. When the ISI team arrived in the field, they established an in-country team, which then reviewed and modified the interview and analysis formats. Finally, the sustainability questionnaire was used to gather relevant data and carry out the assessment.

Several difficulties were encountered in the investigation phase. In many instances the scope of work required time-consuming adaptation to add the distinctive action-inquiry perspective, plus extra negotiations once the teams were in the field. Consultants had to be trained in the use of the questionnaire. However, it is noteworthy that field responses to the assessments were very positive. Clients were pleased, willing to cover local costs, and interested in arranging future field work.

The third phase of the action stream involves the design of alternatives: recommending policy, structural and management changes to improve sustainability. The team used the SCOPE framework, the assessment data, and suggestions from local stakeholders to recommend changes. The ISI team members tried to act as resource persons and facilitators in this process. From the interim assessment and dialogue with interested parties, the team developed a final assessment and made recommendations to USAID and developing country officials.

In step four of AIM, sustainability recommendations are accepted by the organization and are implemented with monitoring of results built into the process. Follow-up visits to several countries indicate that many recommendations have been agreed to and some implemented. However, to date, no formal mechanism exists to track this dimension. One of the current ISI activities is the design of such a mechanism to follow up the work of the first three action stream phases.

Finally, in knowledge consolidation and dissemination, evaluations will continue to be conducted in the field and assessments will be fed back into the research stream. For ISI this would mean that additional sustainability evaluations will be done in the field, and assessments of the utility of ISI action guidelines will be made. In fact, time is needed to develop the guidelines and assess their utility from a sustainability perspective. Currently there are neither mechanisms nor resources for this type of systematic evaluation.

The Inquiry Stream

The preparatory phase in the research component of AIM began with a review and synthesis of previous findings on sustainability, leading to the development of the SCOPE framework and the selection of indicators for its variables. It is important to point out that converting a conceptual framework into indicators and research hypotheses normally leads to some revision of the framework. Moving from theory to research design requires further clarification of concepts and relationships.

Phase two in the inquiry stream consisted of planning the investigation, the field testing of the SCOPE framework, and its application to an existing data set. The major activity in planning the research was the construction of an information questionnaire for the field team, incorporating the SCOPE variables. The logistics and politics of the seven field applications led to the decision that only the Washington-based social scientist members of the ISI teams filled out the questionnaire in six of these cases.

Team members conducting field inquiries were experienced researchers and interviewers; they determined their own research tactics and methods. Invariably they interviewed key project personnel and stakeholders and reviewed previous studies and documents on the project. Their knowledge was converted into a form that was comparable across cases when they completed the information questionnaire on their project or institution. They also reported on the aspects of SCOPE that were most useful for analyzing sustainability issues in the particular case, and what was not useful. Furthermore, they discussed errors and weaknesses in the application of the SCOPE hypotheses to the case.

The SCOPE framework has also been applied to a set of 25 agricultural education institutions. Twenty-five expert questionnaires have been statistically analyzed, and the results support all of the SCOPE propositions, though not always at statistically significant levels. In the design of the inquiry stream, the team compiled and analyzed the data from the field studies and data set studies, and revised the SCOPE framework accordingly. The field cases have not been selected randomly but are the total set of projects for which a request was made for a sustainability analysis. It may be assumed that the on-site interest in action surpassed the research interest in the selection of cases.

This means that the sample parameters cannot be used as estimates of the population parameters. It also means that correlation findings cannot be generalized uncritically. However, unless the cases are rather peculiar, the correlation analysis can be cautiously used as a guide until more representative cases can be analyzed. The main precaution to take in using the findings is to note the unique features of the set of cases. Two such features stand out. Since each project requested a sustainability analysis, these cases probably have sustainability problems and also have project personnel who are concerned about them. Second, six of the seven cases studied to date involve agricultural education or research institutions. (The seventh involves the institutionalization of a credit programme.) The institutional homogeneity of the case limits the generalizability of the correlation results.

Preliminary findings from the data set study of 25 agricultural education institutions indicate that most management factors have surprisingly weak correlations with prognoses of sustainability. Relations with the minister of agriculture seem to override most other contributing factors. It is hoped that further analysis will ferret out the contingency factors that determine which aspects of the SCOPE framework are critical in different conditions.

The implementation phase of the research component of AIM involves research on the implementation of changes in the studied projects, and the development of guidelines for designing and implementing projects for sustainability for donor and country use. ISI has not reached this phase or the final phase — when the team returns to the field and analyzes the results of the implemented sustainability changes, revises the SCOPE framework and the guidelines, and disseminates the findings.

AGRICULTURE AND THE RURAL COMMUNITY

Perhaps in no other area of the "new sociology of agriculture" is there so much apparent continuity with the pre-Korean War tradition of rural sociological research as in research on farm structure and rural communities. The early rural sociologists primarily conceived of agriculture as one of the major institutional ensembles affecting rural community life and, especially during the 1930s, devoted much of their research effort relating to agriculture to understanding the implications of agricultural structural changes for rural community life and viability.

The conceptualization of agriculture and rural community interactions and the methodologies for exploring these interactions in pre-Korean War and post-1970 rural sociological research, however, exhibit dramatic differences. Whereas most of the early rural sociologists tended to be less interested in agriculture per se and were principally concerned with understanding rural community institutions, contemporary researchers have tended to be more interested in structural change in agriculture and have deemphasized understanding the totality of the fabric of rural social life.

The research tools and data sources of these two generations of rural sociological researchers are also dramatically different. The early rural sociologists relied on rural social surveys, while the bulk of contemporary research is based on county-level census data. A further difference between the early and contemporary rural sociologists of agriculture in their work relating to rural communities concerns the pivot of the latter studies around Goldschmidt's "As You Sow". This is somewhat ironic in that Goldschmidt's study was anthropological and qualitative, whereas the research conducted since 1970 that has drawn on Goldschmidt's work has largely been quantitative and has involved far less "anthropological" detail on rural communities than did Goldschmidt's.

Rodefeld is generally credited with the "rediscovery" and popularization of Goldschmidt's pioneering work among rural sociologists, though Heffernan (1972) made important contributions to elevating "Goldschmidt-style" research to a prominent position in rural sociology. Nonetheless, after Rodefeld's refocusing of rural sociological attention on the Goldschmidt thesis, Goldschmidt's book has been virtually an obligatory citation in research articles on this topic. Despite some trenchant criticisms of Goldschmidt's perspective and research methodology, Goldschmidt's general hypotheses remain accepted theoretically and thus continue to shape, if not dominate, research on farm structure and rural communities within the "new sociology of agriculture."

Briefly, Goldschmidt's thesis was that there are differences in the quality of rural community life depending upon the social organization of agriculture and its attendant occupational structure. Goldschmidt observed that the quality of rural life in two carefully matched California communities in the San Joaquin Valley differed dramatically and that these differences were largely accounted for by the organization of agriculture. He argued that Dinuba, a community surrounded by "family" farms, had a greater number of businesses, a larger retail sales volume, a higher per capita income, and a greater diversity of social, educational, recreational, and cultural institutions than did Arvin, a community surrounded by large, industrial, corporate farms. These differences were principally accounted for by an intervening variable — community occupational composition — that was seen to be strongly influenced by the social organization of agriculture and which directly contributed to the dramatic variations in indicators of the quality of life. Goldschmidt reported that the high ratio of laborers to other occupational groups in Arvin accounted for its low level of community viability relative to Dinuba. From these observations Goldschmidt inferred that the rise of large-scale corporate agriculture then underway in California and elsewhere would have adverse implications for rural communities.

Goldschmidt's study received essentially no further attention until it was referred to in a 1968 hearing on corporation farming before the Senate Select Committee on Small Business. LaRose and Sonka and Heady soon reported further data consistent with the Goldschmidt study. In addition to Rodefeld's and Heffernan's efforts to resurrect Goldschmidt's work in rural sociological circles, there was a major research effort by the Small Farm Viability Project in California, drawing largely on Isao Fujimoto's unpublished research, which presented a larger restudy of the Goldschmidt thesis. The results largely supported Goldschmidt's findings of over three decades earlier. Flora and Conboy also reported results consistent with the thesis.

Since that time there have been numerous studies in the new sociology of agriculture tradition that have explored the Goldschmidt thesis in a number of contexts. Several observations can be made on the conduct of recent research in the Goldschmidt tradition, what we might call "neoGoldschmidt" research. First, it must be stressed that contemporary rural sociological researchers have taken considerable liberties with Goldschmidt's formulations. As noted earlier, Goldschmidt's method was a comparative anthropological case study of two communities and eschewed quantitative analysis of the sort that has predominated in the subsequent rural sociological literature.

Also, Goldschmidt's analysis focused on several aspects of the property relations and control over labor, such as absentee ownership and use of ethnic minority laborers, rather than farm size, which has been the most important farm structural indicator in the neo-Goldschmidt tradition. In that regard it is important to stress that the "family" farms of Dinuba were by no means "small" farms; in fact, the family farms of Dinuba were generally larger in gross sales terms than most large farms in the Midwest today.

Second, neo- Goldschmidt research has largely focused on a single direction of causality — the impact of agricultural structure on rural communities — and has largely ignored other types of interrelations between agriculture and rural communities. Third, most studies have been based on Census of Agriculture data and have employed countylevel data as a proxy for community characteristics have utilized survey data to infer community quality of life consequences of changing agricultural structures. Fourth, despite obvious major regional variations in farm structures and in rural communities and their economies, there has, until very recently, been a virtual lack of a comparative perspective on farm and community structures.

Goldschmidt and Harris and Gilbert are partial exceptions to this generalization, since they have done research using states as the unit of analysis, though this procedure compounds the aggregation bias problems. Very recently, nonetheless, there has appeared a series of regionally based studies of the effects of farm structure on rural community viability and quality of life commissioned by the Office of Technology Assessment. These studies are Buttel et al., Skees and Swanson, Flora and Flora, MacCannell, and van Es et al. as published in abridged form in Swanson. These five studies were very briefly summarized in the main report. Together they indicate strongly that there are pronounced regional variations in how changing farm structures affect rural communities.

The impacts of farm structural changes on rural communities are generally most pronounced in the Great Plains and industrial agriculture states, less so in the South and Midwest, and very little in the Northeast. There is evidence that the degree to which farm structure affects rural communities is directly associated with the dependence of the rural community on agriculture.

The general model pursued in these neo- Goldschmidt studies has been to estimate coefficients for a causal chain in which farm structural change indicators are seen to affect the composition of the farm population (*e.g.*, as measured by indicators of the size of the farm population, the proportion of full-time hired workers) and the size and composition of the rural population. These clusters of variables are then seen to affect indicators of the quality of life in rural-agricultural communities (*e.g.*, proportion of families in poverty, median family income, per capital retail sales, unemployment rate).

This general model is sometimes supplemented by technological factors as antecedents of farm structure, since technological change is considered to be one of the major factors influencing farm structural change. More recently, studies in this genre have gone beyond the cross-sectional census data designs that predominated from the mid-1970s to the early 1980s and have included measures of change, where appropriate, in variables in the model. One limit to studies that use census and other data based on single-year assessments of farm income is that net farm incomes are highly variable over the short-run; for example, as many as 20 percent of farms with annual sales of over $250,000 had poverty-level incomes in 1986.

Heffernan has effectively summarized the general configuration of results in the neo-Goldschmidt literature by noting that "all relevant research to date suggests that a corporate type of agriculture results in a reduction in the quality of community life for at least some people, especially the hired workers in rural communities." As such, the results continue to lend support to the Goldschmidt thesis. But there have recently been several expressions of dissatisfaction with Goldschmidt's perspective and with the generalizations that have been made from research in this tradition.

A major theoretical concern is that Goldschmidt and many researchers working in this tradition have taken large-scale and corporate agriculture to be basically synonymous when, in fact, not all large-scale farms are corporate or industrial-type farms. Much more important than industrial-type farms in farm numbers and shares of sales, assets, and profits are larger-than-family farms. Empirically, as Goldschmidt-style research has been done in the Midwest and other traditional family farming areas, there have been admonitions that large-scale agriculture, especially of a nonindustrial character, does not necessarily lead to adverse impacts on rural communities, though some studies note that increased scale of production could have detrimental impacts on community services.

A further area of dissatisfaction with research on farm structure and rural communities is that the treatment of farm structure in this literature has generally failed to incorporate new concepts relating to agricultural dualism. Typically farm size or scale is operationalized as a single variable representing a central tendency measure of farm size, which was a suitable procedure when there was a singular process of differentiation among farms. But with the emerging dualism in U.S. farm structure since the late 1970s, there has been little change in mean acres per farm since the 1969 census. This measure of central tendency thus ignores the processes of dualism that have contributed to increased concentration with apparent stability in average acres per farm since the late 1960s.

The neo- Goldschmidt conceptualization of an agricultural structure in which particular communities can be characterized along a continuum of farm sizes ranging from small to large has also been questioned by Wimberley who, using multiple indicators from county-level Census of Agriculture data and a longitudinal design, found that there are instead three continua relating to degree of "corporate-commercial" agriculture, "largefarm-area" agriculture, and "small-farm" agriculture. Using Wimberley's measures and including sources of off-farm employment as well, Reif found that large-farm-area agriculture and core-industry structures were related both to family poverty and income inequality in counties.

Thus, while studies of the impacts of changing farm structures on rural communities have been a productive area of research in the new sociology agriculture, a number of research frontiers for the future are suggested by recent criticism. There is a need to diversify the methodological approach, consider alternative directions of causality, introduce more sophistication in terms of the measurement of change in farm structure, and expand recent efforts at comparative regional and interdisciplinary research.

The work of Vail represents an encouraging thrust in this direction. Vail has conducted extensive research on the relationships between agriculture and rural communities in Maine in which a variety of factors of demonstrated importance — the reciprocal relations between agriculture and rural communities, the role of an increasingly dualistic agrarian structure, nonfarm aspects of agriculture, and the historically specific nature of farm-community interactions — have been incorporated into a single study.

The Farm Crisis of the 1980s

The notion that American agriculture is no longer (if it ever was) isolated from the remainder of the political economy was driven home clearly by the farm crisis of the 1980s. This crisis was brought by the conjunction of a number of factors in the 1970s, including: accelerated demand for grain on world markets, public policies that encouraged massive capital investments and expansion of production, and a political-economic context that encouraged land speculation. The farm crisis emerged in the 1980s for a number of reasons. Global agricultural export markets shrank, in large part due to the global recession. Farm commodity prices and ultimately net farm incomes (exclusive of federal commodity programme payments and price supports) declined due to the glutting of world grain markets. Reagan administration fiscal and monetary policies led to high real interest rates, which were especially detrimental to agriculture because of its interest rate sensitivity; accordingly, debt-asset ratios among farmers increased rapidly.

The 1980s also witnessed the emergence of new Third World agricultural export competitors (*e.g.*, Argentina, Brazil) because of the global debt crisis and these countries' need to increase exports in order to repay foreign loans. Further, federal fiscal austerity eroded the federal government's ability to maintain farm incomes through commodity programmes (even though commodity programme expenditures soared during the decade). The result was a massive decapitalization of American agriculture, especially in the Midwest and Great Plains, as farmland prices in the states most affected by the crisis declined by over 50 percent.

This crisis attracted a good deal of attention by the mass media and by agricultural economists. It was not, however, until the middle of the decade that much was published about the farm crisis by rural sociologists. Most of this research can be divided into four categories: (1) neo-Marxian (and related neo-Weberian) analyses of the continuing penetration of capitalism into and the role of the state in agriculture, (2) analyses of the differences in the characteristics between farm(er)s that were and were not experiencing "financial stress," (3) studies of the effects of farm financial crisis and failure on the farm families whose enterprises were affected, and (4) studies of public reactions to the farm crisis.

The neo-Marxian (and neo-Weberian) approaches have largely focused on understanding continuing cycles of crises in agriculture under capitalism.

Although these analyses focus heavily on the political and economic context of agriculture, they are not fundamentally different from what is discussed and therefore are not discussed specifically here. Examples of this genre specifically focused on the crisis of the 1980s include Berlan, Bonanno, Mooney, McIntosh and Zey-Ferrell, Zey-Ferrell and McIntosh, Goodman and Redclift, and Buttel. We also briefly examine the farm crisis from a methodological angle — namely, the growing tendency in the political economy of agriculture to stress periods of agricultural crisis as key "test cases" for assessing theories of agrarian transition.

Studies in the second category, analyses of the characteristics of farms with various levels of "farm financial stress," deal more with the effects of the political and economic environment of agriculture on farms of different characteristics. These studies have produced inconsistent results due to regional, ethnic, and other variations within the farming population.

For example, farm-size variables were found to be directly correlated with financial difficulty on Iowa farms and unrelated for Texas and North Dakota farms. Yet aggregate national data revealed that larger farms were more profitable, a relationship that obscured the underlying failure of many middle size farms. Farm families with fewer years of farming and younger ages were affected more severely.

However, Murdock et al. found the cohort that had started farming between the end of World War II and the mid1960s also had higher rates of financial difficulties, suggesting a life-cycle effect for those farm families bringing another generation into the farm. The literature does, however, show agreement on the relationship between other variables and financial crisis. For example, higher debt-to-asset ratios brought by low commodity prices in a time of deflation of asset values were associated with financial problems.

The third category, analysis of the effects of the farm financial crisis and failure on the farm families whose enterprises were affected directly, has largely been the work of William and Judith Heffernan, who have conducted in-depth interviews among Missouri farm families who were in severe financial crisis or had been displaced from agriculture. These families experienced a host of difficulties beyond the simple loss of their farms and the disorientation that accompanies the loss of a valued way of life. Among their major results are the following: Many farm families felt socially isolated and were not well-prepared for participation in nonfarm labor markets.

Local employment opportunities were often unavailable, and tax laws devastated some families economically after they had attempted to leave farming with some resources with which to start a new life. While the long-term effects of farm failure for these families were not yet known, it was clear that some families were adapting to their new situations successfully while others were at risk of suffering long-term poverty. Graham's study of displaced farm families in New York State bears out these findings. This category of studies has implications for the mutual effects of agriculture and its environment as the displaced farm families affect local labor markets and community social structures.

The final category of research on the 1980s farm crisis, on public reactions to the crisis, is best represented by Lyson's research on public apathy. Drawing on theories of public opinion formation, Lyson demonstrates that several factors led to a surprisingly high degree of apathy by nonfarm people towards the farm crisis, despite the massive media attention the crisis attracted. He notes that there has been a trend towards farmers and consumers having very little direct contact and that the relationship between food prices and farm income has become concealed. Most importantly, Americans have come to view "bad news" from the farm sector — the cost-price squeeze, high debt loads, and the loss of family farms — as being normal. Thus the American public, though "interested and concerned" about the economic woes of farmers, exhibited apathy towards the farm crisis.

A major neglected area in the rural sociology literature on the farm crisis to date is studies of the effects of the farm crisis on rural communities. This topic pertains, in particular, to how conditions in agriculture affect its environment, especially at the local level. Financial losses to local agribusiness firms when farms fail, losses of farm families in local rural market areas, declining property tax revenues due to deflated farm asset values, and other problems associated with the declining prosperity of agriculture have likely had severe effects on local services and on the local social fabric.

AGRICULTURE AND RURAL DEVELOPMENT

A SAGA OF SUCCESS

From a nation dependent on food imports to feed its population, India today is not only self-sufficient in grain production but also has a substantial reserve. The progress made by agriculture in the last four decades has been one of the biggest success stories of free India. Agriculture and allied activities constitute the single largest contributor to the Gross Domestic Product, almost 33 percent of it. Agriculture is the means of livelihood of about two-thirds of the workforce in the country.

This increase in agricultural production has been brought about by bringing additional area under cultivation, extension of irrigation facilities, the use of improved high-yielding variety of seeds, better techniques evolved through agricultural research, water management, and plant protection through judicious use of fertilisers, pesticides and cropping practices.

CROPS

The 1970s saw a multi-fold increase in wheat production that heralded the Green Revolution. In the next decade rice production rose significantly;

in 1995-96, rice production was 79.6 million tonnes. The total grain production crossed 211 million tonnes in 2001-02, a big leap from 51 million tonnes in 1950-51.

To carry improved technologies to farmers, a National Pulse Development Programme, covering 13 states, was launched in 1986. Efforts to boost pulse production were augmented further by the Special Food Production Programme. In 2001-02, pulse production was 13.52 million tonnes. With some states offering more than the statutory minimum price, sugar cane production also received a boost, and in 2001-02 a record 292.2 million tonnes was registered.

IRRIGATION

As efforts continued to increase the irrigation potential in the country, the last 40 years saw the gross irrigated area reach 85 million hectares. Flood forecasting has become an important activity over the years. Over 500 hydrological stations collect and transmit data through 400 wireless stations for issuing forecasts for 157 sites. About 5,000 forecasts are issued in a year with 94 percent accuracy. The country also receives international support, with the World Bank as a primary source, for developing its water resources. International cooperation is also envisaged in setting up a National Centre for Information on Water and Power. As there is a broad seismic belt in the country, particularly along the Himalayan, and the Kutch region and parts of Maharashtra, a scheme is being evolved to collect all data on seismic activity at various dam sites.

FERTILISERS

The fertiliser industry in India has grown tremendously in the last 30 years. The Government is keen to see that fertiliser reaches the farmers in the remote and hilly areas. It has been decided to decontrol the prices, distribution and movement of phosphatic and potassic fertilisers. Steps have been taken to ensure an increase in the supply of non-chemical fertilisers at reasonable prices. There are 66 fertiliser quality control laboratories in the country. Since biofertilisers are regarded as an effective, cheap and renewable supplement to chemical fertilisers, the Government is implementing a National Project on Development and Use of Bio-fertilisers. Under this scheme, one national and six regional centres for organising training, demonstrating programmes and quality testing of bio-fertilisers has been taken up.

It was a challenging decision of the Government to take Bombay High gas through a 1,700 km pipeline to feed fertiliser plants located in the consumption centres of North India. However, the major policy which has ensured the growth of the fertiliser industry is the thrust on accelerating fertiliser consumption by fixing, on the one hand, low and uniform price for fertilisers, and on the other hand providing the manufacturers adequate compensation through the retention price and subsidy scheme. As expected, fertiliser nutrient demand has gone up from 0.29 million tonnes in 1960-61 to 16.7 million tonnes at the end of 2000-01, compared to 12.15 million tonnes during 1992-93.

FISHERIES

Fish production achieved an all-time high of 5.6 million tonnes at the end of 2001-02. Programmes that have helped boost production include the National Programme of Developing Fish Seeds, Fish Farmers' Development Agencies and Brackish Water Fish Farmers' Development Agencies. The Central Institute of Fisheries Nautical and Engineering Training trains the necessary manpower. To diversify fishing methods and introduce processed fish products on a semi-commercial scale, an Integrated Fisheries Project has been launched. A National Fisheries Advisory Board has also been established.

FOOD PROCESSING

The Ministry of Food Processing Industries, set up in July 1988, is the central agency of the Government responsible for developing a strong and vibrant food-processing sector with a view to create increased job opportunities in rural areas, enable the farmers to reap benefit from modern technology, create surplus for exports and stimulating demand for processed food. A new seeds policy has been adopted to provide access to high-quality seeds and plant material for vegetables, fruit, flowers, oilseeds and pulses, without in any way compromising quarantine conditions. Initiatives have been taken to encourage private sector investment in the food processing industry.

AGRICULTURAL RESEARCH

The apex body for education, research, training and transfer of technology in the field of agriculture is the Indian Council of Agricultural Research (ICAR), established in 1929. India's transformation from a food deficit to a food surplus country is largely due to ICAR's smooth and rapid transfer of farm technology from the laboratory to the land. ICAR discharges its responsibilities through 43 research institutes, four national research bureaux, 20 national research centres, nine project directorates, 70 all-India coordinated research projects, and 109 Krishi Vigyan Kendras (farm science centres). Besides, the programme of Agricultural Education is coordinated by ICAR with the curricula and other normative guidance given to the 26 agricultural universities and four national research institutes.

OILSEEDS PRODUCTION

A Technology Mission on Oilseeds was launched in 1986 to increase production of oilseeds in the country and attain self-sufficiency. Pulses were brought under the Technology Mission in 1990. After the setting up of the

Technology Mission, there has been consistent improvement in the production of oilseeds. The oilseed production, which was 108.3 lakh tonnes in 1985-86, has increased to 247.30 lakh tonnes in 1998-99, the highest production so far. However, due to unfavourable weather conditions prevailing in the major oilseeds-growing states, the oilseed production during 1999-2000 and 2000-01 suffered some decline.

The increase in production was largely contributed by soyabean, rapeseed and mustard. Production of pulses has seen many ups and downs, which is expected to be checked under the mission. The country grows mainly nine oilseeds, with groundnut, rapeseed and mustard accounting for 62 percent of total production. Lately, soyabean and sunflower have shown major growth potential.

DRINKING WATER

A Technology Mission on Drinking Water and Related Water Management has been constituted to cover the residual problem villages and provide potable water at 40 litres per capita per day, and 70 litres per capita per day in desert areas inclusive of 30 litres for cattle. The mission is tackling the problem through 55 mini-missions in project districts and countrywide problem-oriented sub-missions. A Village Level Operation and Maintenance (VLOM) pump called India Mark-II has been developed and is being exported to 40 countries.

AGRICULTURE IN INDIA

Agriculture in India is one of the most prominent sectors in its economy. Agriculture and allied sectors like forestry, logging and fishing accounted for 18.6% of the GDP in 2005 and employed 60% of the country's population. It accounts for 8.56% of India's exports. About 43% of India's geographical area is used for agricultural activity.

Despite a steady decline of its share in the GDP, agriculture is still the largest economic sector and plays a significant role in the overall socio-economic development of India. The monsoons play a critical role in the Indian sub-continent's agriculture in determining whether the harvest will be bountiful, average, or poor in any given year. The entire rainfall in the sub-continent is concentrated in the few monsoon months.

Agriculture in India is constitutionally the responsibility of the states rather than the central government. The central government's role is in formulating policy and providing financial resources for agriculture.

The government administers prices of essential commodities to protect farmer's interests. It also administers other commodities which are produced by government-controlled companies, like petroleum, coal, nitrogenous fertilizers, etc. Other than these, most agricultural commodity markets operate under the normal forces of demand and supply.

HISTORY OF INDIAN AGRICULTURE

Evidence of the presence of wheat and some legumes in the 6th millennium BC have been found in the Indus Valley. Oranges were cultivated in the same millennium. The crops grown in the valley around 4000 BC were typically wheat, peas, sesame seed, barley, dates and mangoes. By 3500 BC cotton growing and cotton textiles were quite advanced in the valley.

By 3000 BC farming of rice had started. Other monsoon crops of importance of the time was cane sugar. By 2500 BC, rice was an important component of the staple diet in Mohenjodaro near the Arabian Sea. The Indus Plain had rich alluvial deposits which came down the Indus River in annual floods. This helped sustain farming that formed basis of the Indus Valley Civilization at Harappa. The people built dams and drainage systems for the crops. By 2000 BC tea, bananas and apples were being cultivated in India. There was coconut trade with East Africa in 200 AD. By 500 AD, egg plants were being cultivated.

AGRICULTURAL MOVEMENTS

Green Revolution

After independence, considering India's growing population, the government took steps to increase the food production. Yields per unit area of all crops have grown since 1950. The 1970s saw a huge increase in India's wheat production. This is known as the Indian Green Revolution. Reasons for the growth are the special emphasis placed on agriculture and steady improvements in irrigation, technology, application of modern agricultural practices and provision of agricultural credit and subsidies

Operation Flood

Operation Flood was the name of a rural development programme started by the National Dairy Development Board (NDDB) in 1970 with the objective of creating a nationwide milk grid. This movement followed the Indian green revolution and helped in alleviating poverty and famine levels from dangerous proportions in India during the era. It resulted in India becoming the largest producer of milk and milk products, so it is also called the White Revolution of India.

PRODUCTION

India is ranked second in the world in terms of agricultural output. India is the largest producer in the world of milk, cashew nuts, coconuts, tea, ginger, turmeric and black pepper. It also has the world's largest cattle population (193 million). India ranks second worldwide in farm output.

It is the second largest producer of wheat, rice, sugar, groundnut and inland fish. It is the third largest producer of tobacco. India accounts for 10 per cent of the world fruit production with first rank in the production of banana and sapota. Given below is a chart of trend of output of cereals and major foodgrains as published by the Department of Food and Public Distribution with figures in tonnes.

Year	Cereals	Rice	Wheat	Coarsegrains	Pulses
2001–02	199,480,000	93,340,000	72,770,000	33,370,000	13,370,000
2004–05	192,730,000	87,800,000	73,030,000	31,880,000	13,670,000

Despite high growth, international comparisons reveal that the average yield in India is generally 30% to 50% of the highest average yield in the world.

PRODUCTIVITY

The following table shows the harvest per acre of farmland in 2007 with figures in tonnes.

Commodity	Output per acre
National Average	0.29
Sugarcane	26.15
Potato	6.32
Wheat	0.84
Rice	0.70
Corn	0.64
Groundnut	0.37
Soyabean	0.32

The low productivity in India is a result of the following factors:

- Illiteracy, reforms and inadequate or inefficient finance and marketing services for farm produce.
- The average size of land holdings is very small (less than 20,000 m^2) and are subject to fragmentation, due to land ceiling acts and in some cases, family disputes. Such small holdings are often over-manned, resulting in disguised unemployment and low productivity of labour.
- Adoption of modern agricultural practices and use of technology is inadequate, hampered by ignorance of such practices, high costs and impracticality in the case of small land holdings.
- Irrigation facilities are inadequate, as revealed by the fact that only 53.6% of the land was irrigated in 2000-01 which result in farmers still being dependent on rainfall, specifically the Monsoon season. A good monsoon results in a robust growth for the economy as a

whole, while a poor monsoon leads to a sluggish growth. Farm credit is regulated by NABARD, which is the statutory apex agent for rural development in the subcontinent.

In the last few decades several farmers have committed suicide especially in the states of Andhra Pradesh, Maharashtra, Karnataka, and Kerala. Combating this has become a major challenge for these governments. Some of the causes for the deaths include indebtedness of small and marginal farmers and repeated crop failures.

WASTE

As per the report of the 11th Planning Commission, preventable post-harvest losses of foodgrains in India were estimated at about 20 million tonnes a year, which is nearly 10.5 percent of the total production due to inadequate storage, security and distribution facilities.

AGRICULTURE IN THE INDIAN STATES

About 70% population of the India is directly dependent on agriculture, also agriculture is the main component for most of the state economies in India. The Punjab region is suited for wheat-growing. The south Indian states of Andhra Pradesh, Tamil Nadu and Karnataka are also major producers of rice. These states are mostly irrigated by the rivers Krishna, Godavari and Cauvery. Haryana is self-sufficient in food production and the second largest contributor to India's central pool of food grains. Haryana contributed significantly to the Green Revolution in India in the 1970s. The high-altitude states of Himachal Pradesh and Jammu and Kashmir are ideal for production of apples. Arunachal Pradesh also has a large number of fruit orchards. Tea is the other produce of the high altitude regions of Assam, West Bengal (Darjeeling), Tripura, Tamil Nadu (Ooty), Himachal Pradesh and Kerala. Assam produces some of the finest and most expensive teas in the world. However Assam tea is grown at elevations near sea level, giving it a malty sweetness and an earthy flavor, as opposed to the more floral aroma of highland (*e.g.*, Darjeeling, Taiwanese)

In the north-eastern states like Arunachal Pradesh and Nagaland, shifting cultivation known locally as jhum, was practised by the tribal groups, but that has come to be less practised. In mountainous states like Sikkim, farming is done on terraced slopes. The state has the highest production and largest cultivated area of cardamom in India. Karnataka is the largest producer of coffee in India and accounts for 59% of the country's coffee production. It is grown mostly in the lower slopes of the Western Ghats in Kodagu district. Karnataka is also the largest producer of sandalwood based products like perfumes. Sandalwood comes from the dense forested areas of southern Karnataka. The state is also the largest producer of raw silk and Mysore silk saris are famous. The other state manufacturing silk is Tamil Nadu. Rajasthan is among the largest producers of edible oils in India and the second largest producer of oilseeds. The state is also the biggest wool-producing state in India. Kerala produces 96% of national output of pepper. It is also the largest producer of spices which include cardamom, vanilla, cinnamon, and nutmeg. Kerala is also the largest producer of natural rubber in India (91%). The other state producing rubber is Tripura.' 30 percent of Mizoram is covered with wild bamboo forests, and accounts for 40 percent of India's 80-million-ton annual bamboo crop. There has been an economic shift towards agriculture for some states like Bihar after the mineral rich areas were carved out to form the new state of Jharkhand.

RESEARCH

The Indian Agricultural Research Institute (IARI) established in 1905, was responsible for the research leading to the "Green revolution" of the 1970s. The Indian Council of Agricultural Research (ICAR) is the apex body in agriculture and related allied fields, including research and education. The Union Minister of Agriculture is the President of the ICAR. The Indian Agricultural Statistics Research Institute develops new techniques for the design of agricultural experiments, analyses data in agriculture, and specializes in statistical techniques for animal and plant breeding.

AGRICULTURE AND FOOD SECURITY

Pre-Independence India suffered repeated famines, drought and food shortages. But following the Green Revolution in the '60s, yields and foodstocks rose manifold. Now, 30 years later, Indian farmers have realised the follies of their tryst with intensive agriculture. Despite 70 per cent of the population being engaged in agriculture and allied activities, declining foodgrain production and access to food remain the two biggest problems confronting the country. Liberalisation has made things worse: commercial crops are eating into the fertile land tracts meant for essential foodgrains. And six years after the World Trade Organisation came into existence, the anticipated gains for India from the trade liberalisation process in agriculture are practically zero.

Agriculture Minister C Subramaniam sat huddled with his senior staff. There was tension in the air. Depending on imported foodgrains coming from 12,000 miles away, particularly when it came in driblets following US President Lyndon Johnson's directive to "teach India a lesson", was humiliating enough. What worried him was that food stocks had reached such a precarious low that they were only sufficient for another two weeks. Still worse, there was nothing in transit. Recalls Subramaniam, "As a last resort, I told my officials and experts to identify the nearest food-carrying ships on the ocean throughout the world. I said we would identify those carrying wheat to other countries and appeal to the US President to divert them to India if other countries could wait for another six to eight weeks."

This was in 1966-67, the critical year of drought when India imported 11 million tonnes of foodgrain. A year earlier, India had imported 10 million tonnes of foodgrain. Around this time, the Paddock brothers, often referred to as 'prophets of doom', concluded in their book, Famine 1975, that by the mid-'70s at least half of India would be led to a slaughter house. Considering that India had a long history of hunger and famine, and the subsequent failure of the government's effort to boost agricultural production in the post-Independence era, not many disagreed with the conclusive analysis drawn by William and Paul Paddock.

After all, for a country literally surviving on a 'ship-to-mouth' existence, there was little hope. Soon after Independence, India had begun to seek food aid and in fact emerged as the biggest food importer of this century. In 1951, India got ad hoc assistance of 2 million tonnes of foodgrain to tide over the crisis arising from severe drought in several parts of the country. This was the first major food security on a significant scale after Independence.

Before 1947, Indian history was replete with famine, drought and food shortages. Between 1770 and 1880, as many as 27 food scarcities and famines were recorded. At least 20 million lives were lost in India in about 20 famines that had struck since 1850.

Much of this loss was because of misplaced colonial policies which aimed to derive maximum economic gain at the cost of human suffering and misery. After the British had created a transport infrastructure in the first half of the 19th century, they began encouraging farmers to grow crops that could be exported.

The boom in export trade accompanied by rising prices made farmers shift to cash crops like cotton, indigo, poppy and sugarcane. The area under foodgrain subsequently shrank. In other words, efforts to improve agriculture in colonial India were directly linked to the needs of British industries.

In the first half of the 20th century, India's foreign trade showed the same colonial pattern with exports mainly comprising foodgrain and cotton, jute, oilseeds, opium and indigo. Against this, imports consisted mainly of consumer goods. Meanwhile, the cultivator's preference for commercial crops continued.

As is happening now, the trend during the pre-Independence years showed that the best lands available were used for the cultivation of commercial crops. The cultivation of foodgrains was relegated, to some extent, to inferior lands, the yield of which is much less even in normal years. The result was a steady decline in foodgrain output till the Second World War. Per capita food availability from domestic production too had declined. As B M Bhatia observes in The Famines of India: "India kept the precarious balance between increasing population and declining food production partly by turning from an exporter into a net importer of foodgrains but largely by the easy, though dangerous, expedient of pushing down the already low per capita cereal consumption to still lower levels."

The big gap between minimum requirement and supply continued throughout the post-War period. Till the 1960s, Indian agriculture was not sufficiently able to meet domestic requirements and the country had to rely on foodgrain imports.

SOWING THE SEEDS OF THE GREEN REVOLUTION

Realising the disastrous consequences of the growing gap between the rates of growth of population and food production, a vigorous 'grow more food' campaign was launched. The key goal of economic planning shifted to attaining self-sufficiency. The First Five-Year Plan, following food imports in 1951, gave maximum importance to the growth of agriculture. Among the measures initiated in the '50s to stimulate foodgrain production were land reforms, irrigation, fertiliser production, strengthening of research and the organisation of a national farm extension service. Consequently, increased harvests in the mid-1950s relieved the economy.

It was around this time that the government adopted the US Land Grant model of agricultural research. Agricultural universities, beginning with the Gobind Ballabh Pant University of Agriculture and Technology in Pantnagar in Uttar Pradesh, were gradually set up in all the states. The Land Grant model entailed linking research with extension, and taking research from the laboratories to the farms through an extension network linked with the university. In 1961, the government launched the Intensive Agriculture District Programme (IADP) to enhance productivity in the irrigated areas. Although the aim was to introduce improved seed along with a package of agronomic practices to take advantage of the available irrigation, the programme did not come up to expectations for want of high-yielding varieties of wheat.

The tall Indian wheat varieties did not provide an economic response to the application of chemical fertilisers and irrigation water, which agricultural scientists were desperately looking for. These had a tendency to fall down or lodge when fertilisers were applied, as a result of which the yields stagnated at less than one tonne per hectare. Even though breeding for improved non-lodging wheat varieties was initiated in the '50s, it did not meet with much success.

The missing link was provided by Dr Norman Borlaug, who was then working with CIMMYT in Mexico. A few dwarf wheat strains of spring wheat that he sent to the Indian Agricultural Research Institute in New Delhi, looked impressive. In the next two to three years, a wide range of dwarf material was tested under the All India Wheat Research Project. In August 1964, the then Professor and Head of Department of Plant Breeding and Genetics at the IARI, Dr M S Swaminathan, who later emerged as the architect of India's Green Revolution, proposed the launching of a National Demonstration Programme in farmers' fields. This would verify the results obtained in research plots and introduce farmers to new opportunities opened up by semi-dwarf varieties for considerably improving the productivity.

Says Swaminathan, "When small farmers, who with the help of scientists organised the National Demonstration Programme, harvested over 5 tonnes of wheat per hectare, its impact on the minds of other farmers was electric. The clamour for seeds began and the area under high-yielding varieties rose from 4 hectares in 1963-64 to over 4 million hectares in 1971-72. A small government programme became a mass movement." Subsequently, the government imported 18,000 tonnes of semi-dwarf wheat from Mexico in 1966 to be distributed in time for immediate sowing by farmers in the northwestern regions.

The seeds of the Green Revolution were, therefore, truly sown in the mid-'60s. What followed next is already part of contemporary history. Wheat production rose to 17 million tonnes in 1968, an increase of 5 million tonnes over the highest of 12 million tonnes harvested in 1964. Indira Gandhi officially recorded the impressive strides in agriculture by releasing a special stamp entitled 'Wheat Revolution' in July 1968. The success of wheat was later replicated in rice. Productivity increases were also recorded in cotton, sugarcane, millets and oilseeds.

MEASURES TO AVOID FAMINE

The famine-avoidance strategy that India launched had all the necessary ingredients for a sustained effort. In 1965, the Agricultural Prices Commission was constituted, with the basic objective of assuring fair prices for farm produce. The Commission, an autonomous body, works out the cost of cultivation for 22 agricultural commodities, and suggests a price to the government that includes profit that should serve as an incentive to grow more. The administered price of wheat, for instance, was raised by 71 per cent between 1972 and 1981. Similarly, the price of rice was jacked up by 81 per cent during the same period. Later, to sustain the tempo of the rice revolution, the relative price structure was moved in its favour. In the early phase of the Green Revolution, the price policy favoured wheat, but after the mid-'70s it tilted in favour of rice. For oilseeds and pulses, the reverse was put into action. Relatively high support prices were announced in the early-'90s to make their cultivation profitable.

Perhaps the most significant aspect of the price support mechanism has been the insulation of farmers against a decline in prices. Wheat production was very high in 1971-72, as a consequence of which prices should have declined by at least 25 per cent. Since the government purchased all the wheat that was offered, farmers were actually paid 7 per cent more. Again, in 1982-83, when wheat production was 14.3 per cent higher than the previous year, the prices should have actually declined by 32.5 per cent. With the government purchasing all the wheat that flowed into the markets, however, the farmers realised 18 per cent more.

The biggest achievement of the Indian food policy, dependent more and more on buffer stocking and operational stock holding, has been the avoidance of famine-like conditions. It was with the basic objective of curbing consumption and ensuring an equitable distribution of available food supplies, especially in the deficit areas and among the poorer strata of society, that the Public Distribution System (PDS) was introduced. PDS supplies were also used for the 'food for work' programme as well as other anti-poverty programmes. At present, PDS covers more than 80 million families and the total foodgrains distributed through a network of 40,000 fair price shops accounts for over one-third of the total trade in foodgrain.

AIMING AT SELF-SUFFICIENCY

The Green Revolution was certainly not a stroke of luck. The resilience was the result of progressive farm policies, determination, sustained efforts of scientists and the vigour of the Indian farmers who readily adopted the improved technology. This is what the Rockefeller Foundation had to say about the magnificent strides taken: "... The speed with which India's farmers and scientists suddenly gave their country an approach to an abundant food supply, has never been duplicated on an equal scale anywhere in the world, including the scientifically sophisticated United States."

Since the early-'70s, the country has avoided famine even during adverse weather conditions, through a carefully designed food security system involving the maintenance of both substantial grain reserves and an extensive PDS. The devastating drought of 1987, rated as the most severe of this century, affected nearly 155 million hectares. In spite of this, no starvation and hunger deaths were reported. This was essentially the result of excellent food management based on foodgrain reserves built largely from homegrown wheat and rice.

From the late-'60s, the rate of growth in food production generally exceeded the rate of growth of population. Despite the continuous growth in population, the per capita availability of foodgrain increased from 157 kg per year in 1955 to 177 kg per year in 1995, an increase of 15.7 per cent. Food production has risen roughly four times since the days of Independence, from 50 million tonnes in 1950-51 to 198 million tonnes in 1996-97. At the same time, the population has risen from 300 million to 960 million. Milk production too has risen to 68 million tonnes, the highest in the world. Food self-sufficiency was achieved in the late-'70s with the maintenance of a sizeable food buffer. Lately, India has begun to export foodgrain. About 5 million tonnes of rice and wheat were exported during the last few years. Skimmed milk powder amounting to about 5,000 tonnes has also been exported.

The increase in foodgrains was, however, disproportionate. The highest rate of growth was achieved in Punjab and Haryana, where, at a compound growth rate of 4.63 per cent, foodgrain production jumped from 7.23 million tonnes in 1964-65 to reach an all-time high of 30.33 million tonnes in 1995-96. Uttar Pradesh recorded a compound growth rate of 3.01 per cent. Production in Maharashtra, Madhya Pradesh, Bihar, Orissa and the northeastern states continues to stagger.

Tamil Nadu and Andhra Pradesh, on the other hand, have recorded significant increases in rice yields, with production rising to 9.17 million tonnes and 11.58 million tonnes, respectively.

The Green Revolution was aimed at the better-endowed regions. For millions of farmers languishing in the drylands, constituting more than 70 per cent of the cultivable lands, it continues to be a futile struggle. Despite emphasis on dryland farming during the past several decades, the scenario still remains grim.

The undulating topography and the irregular rainfall pattern have combined to aggravate the situation. That the drylands produce about 42 per cent of the country's food shows that the future of farming lies in these areas. Nearly 83 per cent of sorghum, 81 per cent of pulses and 90 per cent of oilseeds grown in the country come from these areas. The poor yields and the fluctuations in production are indications of the scant attention drylands have received from policymakers and planners.

The problem of increasing productivity on drylands has serious socio-economic implications. With every passing year, the gap between the farmer's yields in irrigated areas and in the dry farming region is widening. One year of drought is enough to push a farmer into a deep well of poverty for another two to three years. Drought is a recurring phenomenon in arid and semi-arid areas. Fifty years after Independence, life for millions of people somehow surviving in the drylands continues to be worse than before.

DECLINE IN PRODUCTION

Thirty years after the dawn of the Green Revolution, Indian farmers are realising that their love affair with intensive agriculture is on the decline. Despite a bountiful monsoon for the tenth year in a row, harvests are not as plentiful as expected. However, with no let-up in the ever-increasing rise in population, Indian agriculture is once again at the crossroads.

Alarm bells have been ringing for quite some time. For nearly a decade, agricultural production had stagnated. The spectacular yield growth recorded in the post-Green Revolution years in Punjab and Haryana has receded into history. Of the multiple problems confronting agriculture, rapid fragmentation of land holdings is keeping pace with increasing population. In 1976-77, the average size of the holdings was estimated at 2 hectares, while in 1980-81, it came down to 1.8 hectares. Today, it stands at a mere 0.2 hectares. The total number of land holdings in 1981 were around 89 million; today these have crossed 100 million. By the turn of the century, the average land holding will come down to 0.11 hectares. It is quite obvious that with such small land holdings, Indian agriculture cannot adopt high-tech farm practices.

Basking in the afterglow of the Green Revolution, farming and agriculture have ceased to attract serious attention. In 1995-96, foodgrain production slumped to minus 3.60 per cent. In 1997-98, it further declined to minus 3.70 per cent, the worst-ever since the heady days of the Green Revolution. Foodgrain production in the frontline agricultural states of Punjab, Haryana and western Uttar Pradesh, comprising the country's food bowl, has decelerated.

The miracle that began with wheat was replicated in rice. This is where the cost-benefit ratio has turned negative. The downward spiral in food production continues through the southern regions of the country. Tamil Nadu, another Green Revolution area, is under tremendous strain from intensive cultivation.

In Karnataka, the negative trend in yield levels of all food crops, barring cotton and sugarcane, are all too apparent. Farming in Karnataka can be clearly separated in two distinct classes, the 'creamy layer' of corporate agriculture occupying the fertile and irrigated areas, and the remaining low productive tracts at the mercy of the rains, constituting the tiny and small land holdings. Kerala too, picturesque and verdant with tropical forests, is passing through its worst-ever crisis on the foodgrain front.

As if this is not enough, the number of landless in rural areas too has been multiplying over the past few decades, at an estimated rate of two million every year. The rapid expansion in the number of land holdings, with less than a hectare under plough, has meant an additional 6,00,000 farmers every five years or so. The negative terms of trade for agriculture and the declining public sector investments in farming are indices of the sluggishness in this sector.

The Economic Survey for 1996-97 has further sounded a warning. The compound growth rate in foodgrain production at 1.7 per cent between 1991 and 1996, is lower than the annual population growth of 1.9 per cent. The survey adds that even a marginal fall of three to four per cent in foodgrain output can cause the prices of essential commodities to escalate sharply, necessitating government intervention by way of resorting to imports. This happened in 1993 and again in 1996.

The philosophy of agricultural planning is changing. Gone are the days when the nation's emphasis was solely on attaining self-sufficiency in foodgrain production. Gone also are the days when farmers were newly independent India's heroes, revered for their role in keeping hunger and starvation at bay. Today, at a time when food production struggles to keep pace with the burgeoning population growth, farmers are being asked to diversify, produce crops that are suitable for export and to compete in the international market.

With the promise of cheap food available off the shelf in the global market, the focus has shifted from agriculture to industry, trade and commerce, from the small and marginal farmers to the agri-processing companies, which alone can bring in investments and add value to produce.

The restructuring of the Indian economy too has put the emphasis on agri-exports. Cultivation of staple food is being replaced by cash crops, tomatoes in place of wheat, durum wheat (for bakery purposes) replacing wheat as a staple diet in Punjab and Haryana, flowers in place of rice, and so on. In the coastal areas, private enterprise is taking away the fish catch, depriving the local communities of a livelihood and their only source of nutrition. In Kerala, vast tracts of forests and paddy fields have been converted into rubber, coffee and coconut plantations. Every year, about 25,000 hectares of good paddy land are being diverted for non-paddy purposes. The structural transformation is not only peculiar to Kerala. It is happening in almost all the states. Commercial crops are eating into the fertile land tracts meant for growing essential foodgrains, thereby making the states rely more heavily on the PDS. The diversion of good agricultural land, which in any case is limited, to commercial farming and even industries, is further restricting the ability to grow enough foodgrain.

IN THE ERA OF WTO

Kerala has already been at the receiving end. Over the past few years, ever since economic liberalisation became the development mantra, the state has been flooded with cheap and highly subsidised agricultural imports, throwing its agrarian economy out of gear. Whether it is the import of palm oil or rubber or coffee, almost every aspect of the state's socio-economy has been negatively impacted.

Coconut prices have crashed, down from Rs 10 to Rs 2. Rubber has plummeted from Rs 60 to Rs 16 and coffee from Rs 58 in 1999 to Rs 30 per kg in 2000. Even spices have not been spared, with pepper prices falling from Rs. 2,600 to Rs. 1,300 per quintal in the consecutive period. Kerala is not alone.

The fallout from the emerging global trade paradigm is being felt all over the country, though not in the same magnitude. Before we go any further however, it is important to understand how the market rules play, and then try to understand the implications for Indian agriculture.

It is now official. Six years after the World Trade Organisation (WTO) came into existence on January 1, 1995, the anticipated gains for India from the trade liberalisation process in agriculture are practically zero. The ministry of agriculture as well as the ministry of commerce have admitted that the hopes from an international regime that talked of establishing a fair and market-oriented agricultural trading system have been belied.

It is time to draw a balance sheet of the gains and losses that accrued from the implementation of the WTO's Agreement on Agriculture, which had incorporated three broad areas of commitments from member states, namely in market access, domestic support and export subsidies. The underlying objective of this was to correct and prevent restrictions and distortions in world agricultural markets.

Six years later, it has been established that these measures have only protected the farmers and the farming systems of the developed countries. On the other hand, the trading regime has ensured that developing countries take time-bound initiatives to open up their domestic markets for cheap and highly subsidised imports of agricultural commodities.

EASIER MARKET ACCESS NEEDED

A recent study by the Food and Agricultural Organisation of the United Nations (FAO) concludes that there has been hardly any change in the volume of exports. Tariff peaks or high import duties continue to block exports from the developing countries.

Tariffs still remain very high, especially in the case of cereals, sugar and dairy products. Sanitary and phytosanitary measures, which were enforced to ensure the quality of imported products, actually continue to be a major barrier in diversifying exports in horticulture and meat products. Selective reduction in tariffs by developed countries has also blocked exports from developing countries. To top it all, only 36 countries (all developed) have the right to impose special safeguards if agricultural imports distort their domestic market!

India was forced to either phase out or eliminate the quantitative restrictions (QRs) on agricultural commodities and products by April 1, 2001. India has, therefore, opened its market. Already, cheaper imports of skimmed milk powder, edible oils, sugar, tea, arecanut, apples, coconut etc have flooded the market.

LACK OF DOMESTIC SUPPORT

Clever manipulation of their subsidy reduction commitments has in reality increased the support to farmers in developed countries. In the United States, subsidy to just 9,00,000 farmers has increased 700 times since 1996. Between 1998-2000, the US has provided an additional US $26 billion to its farmers. In absolute terms, farm support in the OECD countries increased by 8 per cent to reach the staggering figure of US $363 billion in 1998.

In India, we are being told that since our Aggregate Measure of Support (AMS), a measure of the subsidies that are provided to agriculture, are negative, we can still raise our subsidies to farmers. In reality, India is committed to doing away with agricultural subsidies under the Structural Adjustment Programme of the World Bank and IMF. In any case, India provides only US $1 billion worth of indirect subsidies to 550 million farmers!

It was anticipated that due to reduction in domestic support in developed countries, cereal production would shift from developed countries to developing countries. Empirical evidence, however, shows that such a trend is not at all visible. Moreover, with such massive subsidies intact, and with the QRs being lifted, India faces the threat of subsidised food imports.

8

Rainfed Agriculture and Poverty Alleviation

POVERTY AND GROWTH OF RAINFED AGRICULTURE

Measuring poverty and relating it to the growth of rainfed agriculture is a complex topic that we treat only briefly here. We refer to existing literature to review various aspects of rural poverty, and we discuss poverty alleviation measures and their strengths and weaknesses. The literature on poverty in India is large, so we cannot do justice to it. The percentage of people falling below the poverty line in different states between the years 1973-74 and 1987-88. Unfortunately this table does not give us any indication of the distribution of poverty between rainfed and irrigated areas or across different rainfed types, but it does show some important trends. In particular, the percentage of rural people who are considered poor by the poverty line indicator has fallen steadily in every state listed.7 This constitutes part of the evidence for what Singh (1990) refers to as the beginning of the "great ascent" out of poverty in rural South Asia.

Rainfed regions are characterized by higher poverty than irrigated regions. Table below shows clearly, for the year 1973, that the percentage of the population below the poverty line falls steadily as irrigated area rises. The district database does not include information on the number of people below the poverty line, so a rigorous analysis of the relationship between irrigated area and percentage of people in poverty is not attempted here. State level data are available, however, and can give us some indication.

More recent empirical work has suggested that agricultural growth has in fact succeeded in reducing poverty levels. Singh (1990) argued that all income groups gained from the green revolution, but said it is "widely accepted" that wealthier groups benefitted more than less wealthy groups.

Hazell and Ramasamy (1991), on the other hand, show evidence that the large income gains associated with the green revolution between 1973-74 and

1983-84 in North Arcot, Tamil Nadu, did benefit landless and small farm households by at least the same proportion as for large farm households. They found that a 4% decrease in total agricultural employment was more than offset by large wage increases and increased off-farm employment opportunities. Also, the lower on-farm employment figures reflected a drop in labor market participation by members of households farming more than one hectare of land rather than the landless or holders of even smaller plots. Larger farms also gained from the green revolution, but less so due to larger cost increases they faced due to rising labor and fertilizer costs. The conflicting findings of Singh, on the one hand, and Hazell and Ramasamy on the other may reflect differences in the rate of adoption of green revolution technology. Hazell and Ramasamy found that larger farmers adopted much more rapidly than small farmers, but after ten years the use of green revolution technologies was scale neutral.

They also showed that smaller farmers were able to retain their land holdings despite their initially inferior position, enabling them to remain in a position to make strong gains after they eventually adopted. Also, Hazell and Ramasamy only studied one small region. Hazell and Ramasamy also found that every additional rupee of value added in agriculture stimulated growth in value added in the nonagricultural sector equivalent to Rs 0.80. This large multiplier effect has important implications for the role of agriculture in stimulating widespread economic development.

It also helps to offset concerns that may be generated by the reports of the small impact of agricultural growth on agricultural employment, cited above. Ravallion and Datt (1994), in a study of household data from 1951 to 1991, drew conclusions similar to those of Hazell and Ramasamy. They found that rural economic growth spread gains evenly and so brought ample gains to the rural poor; rural growth even helped reduce urban poverty to a certain extent. A related debate addresses the question of whether urban or rural economic growth has the greater impact on poverty alleviation.

Ravallion and Datt (1994) found that urban economic growth was associated with unequal distribution of benefits, so that its impact on urban poverty alleviation was small and the impact on rural poverty was insignificant. This contrasts sharply with their findings above regarding the favorable impact of rural economic growth on poverty alleviation. Ravallion and Datt concluded that both continued rural economic growth — including growth in rainfed agriculture — and a more equitable process of urban growth will be important to poverty alleviation in the future.

INSTABILITY OF RAINFED VS. IRRIGATED AGRICULTURE

Agricultural yields are more unstable in rainfed areas than where irrigation is assured. But if irrigation water supplies fail, then irrigated agricultural area can fall sharply, leading to even more unstable output than under rainfed agriculture. Dhawan (1988a) used the coefficient of variation10 (CV) as the indicator of instability of production, yield and area cultivated for irrigated and rainfed food grains. Using national data, Dhawan found that the CV of output of unirrigated food grains for the period 1970-83 was 11.4%, significantly higher then the 6.4% for irrigated food grains. Decomposing these into output and yield instability, the CVs of yield were 9.2% and 5.3% for rainfed and irrigated food grains, respectively, and for area they were 3.1% and 2.3%, respectively. This suggests that yield and area moved together to increase output instability. Shah and Sah (1993) also found higher yield variations in rainfed than irrigated crops. They calculated CVs ranging from 36% to 70% for yields of rainfed food grain crops and 9%- 26% for irrigated food grains. They did not examine how these yield variations translated into production instability. The larger variations found by Shah and Sah most likely result from their small sample size of a few rainfed and a few irrigated districts over 10 years in the state of Gujarat. Dhawan's study was based on nationwide data, which is likely to be more smooth. Walker (1989b) found similarly large fluctuations in the yields of both rainfed and irrigated crops in a village level study in Maharashtra and Andhra Pradesh. This study was based on 40 households in each of three villages over 10 years, so it is not expected to be smooth. For irrigated paddy in one Andhra Pradesh village, the mean CV was 31% between 1975-76 and 1983-84, and for several rainfed crops in three villages in Andhra Pradesh and Maharashtra, the mean CVs ranged from 44% to 69%. In tank irrigated areas of south India, Hazell and Ramasamy (1991) found that paddy production fell 50% in their study villages in Tamil Nadu in the 1982-83 drought; virtually all of the decline resulted from reduced area planted. Bidinger et al (1990) found an even greater decline in paddy area in their study village in Andhra Pradesh in the 1985-87 drought, with little change in irrigated yields.

INSTABILITY ASSOCIATED WITH IMPROVED AGRICULTURAL TECHNOLOGY

There has been much concern over the possibility that improved crop varieties are associated with increased output instability in addition to higher yields. If this is the case, it could deter adoption of more productive technology and hence retard agricultural development.

INSTABILITY AT THE FARM LEVEL

In discussions of farmers' adoption of HYVs, it is often assumed that new seeds have more volatile yields. Indigenous landraces often have the characteristic of being fairly unresponsive to improved management and increased inputs, but also robust in the face of unfavorable agroclimatic conditions, such as drought. HYVs, on the other hand, often are characterized as being highly responsive to management and inputs, but very low yielding in the event of bad weather. If this is the case, then plant breeders must breed on the basis of one strategy (high average yield) for high potential areas and another (risk minimization) for low potential areas.

Research on pearl millet, which is commonly grown in unfavorable agroclimatic areas, suggests that the above characterization of traditional varieties and HYVs is not necessarily correct. Witcombe (1989), in a study of Pakistan and India, found that good performance of a particular millet seed over all environments appears to indicate good performance in environments of low potential. In three years out of his four year study, the highest yielding entry across all environments was also one of the two highest yielding entries in the lowest yielding environment. These results suggest that the typical plant breeder's strategy of selecting among the highest yielding seeds across all environments is satisfactory. Farmers who adopt HYVs do not necessarily subject themselves to greater risk of catastrophic loss in the event of bad weather.

Farmers' Strategies to Reduce and Cope with Risk. Walker and Jodha (1986) point out that dryland farmers have various methods to reduce their exposure to crop production risk. Cultural practices play an important risk-reducing role; they include planting different crops with relatively low covariate yield (either in an intercrop or on separate fields); diversifying spatially by operating multiple plots with different environmental characteristics; and staggering planting dates in the face of variable rainfall patterns. Sharecropping is a common tenancy arrangement that distributes risk between the tenant and landlord. Many farmers have multiple sources of income, reducing risk if they have low covariation.

Farmers also have various mechanisms to cope with risk that they cannot eliminate. For example, they can borrow from local stores or money lenders, draw down food stocks or savings, sell assets, obtain transfers from relatives, participate in government relief programmes, or migrate. Most of these options are not particularly desirable; for especially poor people they can be quite devastating: selling assets or going into debt may make a family permanently worse off even after drought is over. Jodha indicates that some families will reduce their food consumption as much as possible before parting with their assets; this has obvious negative short term health implications that are particularly severe for those who consume only minimum requirements to begin with.

Effect of Risk on Technology Adoption. While weather-related risk undoubtedly presents great hardship for a very large number of people, Walker indicated that it may not be as important as generally believed in adoption of new technologies in rainfed areas. The tradeoff between expected income and the variance of that income suggests that given farmers' measured risk preferences, the overall effect on adoption decisions is modest. Also, Walker stressed that yield instability does not translate into major variability in income due to farmers' mechanisms to absorb risk. The multiple sources of income and diverse cropping patterns mean that yield variability of one crop only affects a portion of the income from crops. Equally important, in many cases farmers can adjust the area under each crop depending on the weather at the start of the season. For example, dryland rabi (postrainy season) sorghum farmers in black soil areas know at the start of the season how much moisture is available and adjust their cropped area accordingly. Similarly, castor farmers in Andhra Pradesh know that pest attacks are more prevalent when the rainy season begins late, so they plant less castor. And farmers with irrigation from tanks or wells know roughly how much water will be available in the postrainy season, so they adjust planted area accordingly. Hazell and Ramasamy (1991), for example, found that sharply reduced paddy production in North Arcot, Tamil Nadu, in the drought year 1982-83 resulted from a fall in area, not yield. Not surprisingly, Walker (1989b) also found that variations in cultivated area This is why SWC programmes are sometimes referred to as "drought-proofing" measures.

exceeded those in yield. He also found that given farmers' diversified farming and livelihood strategies, the large variations in yield translated into only small variations in income. The variations were so small that even if plant breeders could develop varieties with perfect stability (zero variation), the contribution to household income stability would amount to less than 1% for most of the crops studied, and a maximum of 2.9% for paddy (which, paradoxically, had the least yield variability to begin with).

Walker's (1989b) results demonstrate the importance of looking beyond yield variability of a single crop to variations in all the crops in a given household, village or region, and beyond crop variation to income variation. From this perspective, yield instability does not appear to be a major determinant of adoption, and hence not a top priority for plant breeders. He pointed out (in his 1989a study) that policies related to international trade and storage between surplus and deficit years can be more cost-effective in coping with increasing yield instability.

Soil and water conservation (SWC) investments also are associated with a variety of risks. First, erosion itself is a matter of risk. For example, some plots may be at risk of productivity loss from continuous, gradual erosion, but others may be more susceptible to significant erosion only in the event of a once-in-five-years or once-in-fifty-years storm. For such lands, in normal years there is no gain from investments to reduce erosion, but in exceptional years the gain — actually the avoided loss — may be quite high. Second, often SWC practices serve to conserve soil moisture as well as reduce erosion. Soil moisture retention is more likely to offer immediate, productivity-increasing benefits than erosion prevention. However, these benefits may exceed the costs only when rainfall is unusually low or unevenly distributed, so that moisture stress constrains productivity.12 In a good rainfall year, on the other hand, short term gross returns to increased moisture retention may be low or zero, and in a very high rainfall year they may even be negative if they lead to waterlogging. Little information is available regarding the impact of risk on Indian farmers' SWC investments. The watershed projects are so heavily subsidized that risk is not a factor. The little research conducted on adoption outside of such projects did not address the issue of risk.

INSTABILITY AT THE AGGREGATE LEVEL

Ramakrishna (1993) used a log function to calculate CVs to compare instability of production, yield and area cultivated for cereals between the pre-green revolution and green revolution periods. Ramakrishna found that for food grains, output was more stable during the green revolution than before. Yields were about equally stable between the two periods, and area became more stable. For other crops, on the other hand, Ramakrishna found that output and yield instability both increased during the green revolution, while area again became more stable. Combining food grains and other crops, output and yield instability increased slightly during the green revolution, while area instability was constant. On the whole, Ramakrishna's data suggest that the green revolution had little impact on agricultural instability.

Hanumantha Rao (1994) took a slightly different approach and found different results. Instead of taking CVs of output levels, he took the standard deviation of annual output growth rates as a measure of instability. He found that the standard deviation of output growth of food grains was 8.1 in 1950-51 but 11.4 between 1968-85, and took this to indicate that output instability increased with the green revolution. He also found that the standard deviation rose from 9.4 in the first decade of the green revolution to 11.8 in the second decade. Hanumantha Rao did not provide detailed figures regarding yield and area instability. His figures do not really offer a good comparison to those of Ramakrishna, because his pre-green revolution figures are based only on a single year. Also, it is difficult to compare the magnitude of Hanumantha Rao's figures (expressed as the standard deviation of output growth rates) to Ramakrishna's (expressed as the CV of the level of output). Since we do not know the growth rates from which Hanumantha Rao calculated the standard deviation, we cannot relate his standard deviations to Ramakrishna's CVs. Hazell (1982) examined cereal crop production between 1954-55 and 1964-65 (before the green revolution) and 1967-68 and 1977-78 (during the green revolution, after the introduction of high yielding varieties).13 He found that the CV of production increased by about 50%, from 0.04 to 0.059 between the two periods. Hazell hypothesized that if the increased instability were due to HYVs, variances in production within states would have to rise. However, he found that changes in yield covariances were much more important than changes in yield variances. Only about 18 percent of the increase in variance of total cereal production resulted from changes in crop production variances; the remaining 82 percent was explained by changes in covariances; interstate covariances within crops contributed 41 percent to the change in variance in total cereal production. As a result, Hazell concluded that HYVs were probably not the primary cause of increased variability. In a later paper, Hazell (1984) suggested that HYVs could possibly affect yield covariances of maize in India and the United States, because the narrower genetic base of improved varieties would make them susceptible to common yield inhibitors such as pests and diseases. This would then be a contributing factor to greater variability in national ceral production.

Walker (1989a) studied changes in yield and output variability of sorghum and pearl millet in India resulting from the spread of HYVs. This study followed Hazell's 1984 study relating the rising covariance of yields between states to the spread of HYVs; it used districtlevel data for the 48 largest sorghum districts and the 40 largest pearl millet districts. Walker found that variability had indeed increased significantly between the period 1956-57 to 1967- 68 and 1968-69 to 1979-80, from 8% to 16% for sorghum and 11% to 34% for pearl millet.

He also found that covariance of yields across districts was by far the most important factor in overall output variance of sorghum and pearl millet. Walker examined several possible causes of increased covariance, including 1) changes in rainfall covariance, 2) changes in irrigated area, and 3) adoption of HYVs. He found evidence that provided weak support for each of these possible sources of covariance. However, he also stressed that the contribution of HYVs to increased instability is dwarfed by their contribution to productivity, so he did not recommend changes in existing breeding strategies that develop HYVs for adoption over a large area.

Walker's (1989a) study helps clarify that numerous factors can contribute to changes in stability. Researchers who find that stability increased or decreased with the green revolution should hesitate before proclaiming that HYVs were the cause. Changes in cropping patterns, fluctuations in weather, and the quality of land on which a particular crop tends to be planted all can have an impact on the variations in yield. More importantly, variations in area may either counteract or reinforce those in yield; it is important to know the composition of output fluctuations between yield and area variations. The studies by Hazell and Walker also highlight the need to distinguish between variability at the regional or national levels from farm or field level risks. Because of the dominance of covariance relations in aggregate production data, variability can increase at the aggregate level even while farm level variability changes little or not at all. Although diets were stable, health problems increased due to a shortage of clean drinking water in the village resulting from the drought.

DROUGHT RISK

The preceding discussion on output instability has underemphasized the risks associated with the specific problem of severe, prolonged drought. When severe drought occurs over a wide area, yield and income risks can be particularly great. In Hazell and Ramasamy's (1991) study of North Arcot district, Tamil Nadu, for example, they found that the 50% decline in paddy production in their study villages was replicated throughout the district and beyond. Incomes fell by 50% on average as even nonfarm incomes fell sharply. Consumption expenditure fell by 50% on average, and diet quality deteriorated. Clearly, the effects of drought were covariate across villages and sectors. This problem of covariate risk is particularly challenging because poor performance of one income source, or one location, cannot necessarily be compensated by better performance in another. As a result, people may face severe hardships in the event of drought.

In the Bidinger et al (1990) study, on the other hand, despite a similar 50% decline in income, dietary intake did not change compared to the pre-drought situation.14 People maintained their dietary intake through increases in temporary migration and consumption credit, and government rice subsidies that kept prices low and stable, enabling people to translate meager incomes into normal diets. The number of people who migrated only increased by about 10%, but they stayed away much longer on average. Interest rates did not show the expected increase as demand for consumption credit rose, even though the bulk of borrowing was done through the informal village credit market rather than the formal banking system, which contributed little. Distress sales of land and livestock were rare, but two families did sell all their assets and move permanently to become urban laborers.

GOVERNMENT INTERVENTIONS TO MANAGE RISK

Severe droughts impact negatively on most rural households simultaneously and are therefore difficult to manage through traditional risk sharing and coping strategies. As a result, government policy may sometimes play an important role in helping farmers manage risk. While some government interventions are well-established, other ideas remain relatively untested. Likewise, even where farmers and rural communities have developed effective mechanisms to manage production risk, it is not clear what the costs of these mechanisms are in terms of reduced production efficiency. Perhaps market-based or government-sponsored alternatives can be introduced that protect farmers from the effects of risk but without requiring diverse income sources and fragmented agricultural holdings. Anderson and Hazell (1994) point out that more information is needed about the costs of these risk-reducing and coping mechanisms. We briefly review government measures to manage risk and discuss some additional possible government approaches. Most of the existing drought management efforts. These are poverty allevation programmes to cope with the effects of drought, namely employment programmes and food subsidies. There is no need to repeat the discussion of these programmes here. Bidinger et al (1990), stress the importance of food subsidies in mitigating the effects of drought in their study, but they also point out that a timely public works programme could have prevented much of the unemployment and debt experienced by laborers in the village.

Crop insurance is provided by the public sector in many countries. The impetus for such programmes often originates in governmental concern about catastrophic risks such as drought, or the desire to reduce the incidence of loan defaults to banks. With few exceptions, the financial performance of public crop insurers has been ruinous (Hazell 1992). To be financially viable without government subsideis, an insurer needs to keep the average value of its annual outgoings—indemnities plus administration costs—below the average value of the premiums it collects from farmers. In practice, many of the larger insurance programmes pay out $2.50 or more for every dollar of premium they collect from farmers. The difference is paid by governments, at costs varying from $10 to $400 per insured hectare. Even at these levels of subsidy, many farmers are still reluctant to purchase insurance. As such, many crop insurance programmes are compulsory, either for all farmers growing specified crops or for those who borrow from agricultural banks. The primary reason for the high cost of public crop insurance schemes is that they invariably attempt to insure risks that are prone to severe moral hazard problems, whereby farmers have incentive to lie about their production levels, or to allow their crops to fail in order to receive insurance payments. These risks include many climate, disease and pest risks that are difficult to quantify and assess, and whose damage can be influenced by armers' management practices. The problem is aggravated by a common practice of insuring "target" yields rather than compensating for actual losses. But this is not the only reason for failure.

Another overwhelming factor is the incentive problem that arises oce the government establishes a pattern of guaranteeing the financial viability of an insurer. If the insurance staff know that any losses will automatically be covered by governmetn, they have little incentive to pursue sound insurance practices when setting premiums and assessing losses. In fact, they may find it profitable to collude with farmers in filing exaggerated or falsified claims. Yet another common reason for failure has been that governments undermine public insurers for political reasons. In Mexico, the total indemnities paid has borne a strong statistical relationship with the electoral cycle, increasing sharply immediately before and during election years, and falling off thereafter. In the USA, the government has repeately undermined the national crop insurer (FCIC) by providing direct assistance to producers in disaster areas. Why should farmers purchase crop insurance against major calamities (including drought) if they know that farm lobbies can usually apply the necessary political pressure to obtain direct assistance for them in times of need at no financial cost? Another reason for their high cost is that crop insurers tend to be too specialized, focusing on specific crops, regions and types of farmers, particularly when the insurance is tied to credit programmes designed to serve particular target groups identified by the government. Without a well-diversified insurance portfolio, crop insurers are susceptible to covariability problems, a face the prospe t of sizable losss in some years. Since public insurers are rarely able to obtain commercial reinsurance or contingent loan arrangements, thsi specialization increasestheir dependence on the government.

Crop insurances is provided by the public sector in many countries. The impetus for such programmes often originates in governmental concern about catastrophic risks such as drought, or the desire to reduce the incidence of loan defaults to banks. With few exceptions, the financial performance of public crop insurers has been ruinous. To be financially viable without government subsidies, an insurer needs to keep the average value of its annual outgoings — indemnities plus administration costs — below the average value of the premiums it collects from farmers. In practice, many of the larger insurance programmes pay out $2.50 or more for every dollar of premium they collect from farmers. The difference is paid by governments, at costs varying from $10 to $400 per insured hectare. Even at these levels of subsidy, many farmers are still reluctant to purchase insurance. As such, many crop-insurance programmes are compulsory, either for all farmers growing specified crops, or for those who borrow from agricultural banks. The primary reason for the high cost of public crop-insurance schemes is that they invariably attempt to insure risks that are prone to severe moral hazard problems whereby farmers have incentive to lie about their production levels, or to allow their crops to fail in order to receive insurance payments. These risks include many climate, disease and pest risks that are difficult to quantify and assess, and whose damage can be influenced by farmers' management practices. The problem is aggravated by a common practice of insuring "target" yields rather than compensating for actual losses. But this is not the only reason for failure.

Another overwhelming factor is the incentive problem that arises once the government establishes a pattern of guaranteeing the financial viability of an insurer. If the insurance staff know that any losses will automatically be covered by government, they have little incentive to pursue sound insurance practices when setting premiums and assessing losses. In fact, they may find it profitable to collude with farmers in filing exaggerated or falsified claims.

Yet another common reason for failure has been that governments undermine public insurers for political reasons. In Mexico, the total indemnities paid has borne a strong statistical relationship with the electoral cycle, increasing sharply immediately before and during election years, and falling off thereafter. In the USA, the government has repeatedly undermined the national crop insurer (FCIC) by providing direct assistance to producers in disaster areas. Why should farmers purchase crop insurance against major calamities (including drought) if they know that farm lobbies can usually apply the necessary political pressure to obtain direct assistance for them in times of need at no financial cost? Another reason for their high cost is that crop insurers tend to be too specialized, focusing on specific crops, regions and types of farmers, particularly when the insurance is tied to credit programmes designed to serve particular target groups identified by the government. Without a well-diversified insurance portfolio, crop insurers are susceptible to covariability problems, and face the prospect of sizable losses in some years. Since public insurers are rarely able to obtain commercial reinsurance or contingent loan arrangements, this specialization increases their dependence on the government. Public crop insurers also tend to have high administration costs. This is partly because they often insure small-scale farmers, but also because crop-insurance work is very seasonal, and the absence of a well-diversified portfolio means that staff and field equipment are underemployed for significant parts of the year.

There is no convincing evidence that public subsidization of crop insurance has been socially beneficial. Indeed, social benefit-cost analyses of the Mexican and Japanese schemes show negligible social returns in relation to their high costs. Nor is there much evidence that it has increased agricultural lending or benefited agricultural banks. In a rare study, Pomareda (1984) compared the performance of insured and uninsured loans in the portfolio of the Agricultural Development Bank (BDA) of Panama. Insured loans had slightly higher and more stable returns than uninsured loans. They were also repaid and cleared from the books closer to their expected duration. But the overall gains to the Bank were modest, and could have been achieved more easily at no cost to the government simply by allowing a 2 percent increase in the interest rate that BDA charged its borrowers. This would also have been cheaper for the borrowers than the premium rates they paid for the compulsory insurance.

In order to overcome the major problems associated with crop insurance, and to substantially reduce its administration costs, several authors have proposed area-based yield insurance. Under this proposal, the crop yield for a *homogeneous region* is insured, and all insured farmers in the region pay the same premium and receive the same indemnity. Indemnities are paid whenever the average yield for the region falls below some critical level irrespective of the actual yields obtained by individual farmers. Premiums are calculated on the basis of year-to year variations in the average yield for the region, and would vary from one homogeneous region to another in accordance with differences in risk levels. This approach reduces moral hazard problems, and hence broadens the range of yield risks that can be viably insured. Moreover, since the premiums and indemnities are identical for all insured farmers in a region, it avoids the *adverse selection* problem. The latter refers to the situation in which farmers facing below-average risk tend to drop out of insurance programmes if they are charged premium rates based on average risk levels but are paid indemnities based on their own losses. Also, by eliminating the need for field inspections and loss assessments, the cost of administering an area-based scheme could be kept very low. Providing farmers pay their premium, it is not really necessary that they even grow the crop that they have insured.

Despite its appeal and its potential scope for reaching small-scale farmers, there are problems with the proposal. First, the insurance will be attractive to individual farmers only if their yields are highly correlated with the average yield for the region. Dandekar (1977) argues the homogenous areas can be defined in India in which inter-farm yield correlations are positive, but their is a growing body of micro-evidence showing that yield correlations between plots within the same village, or even within the same farm, are surprisingly low (*e.g.* Walker and Jodha, 1986). Small differences in ground contours, slope, and wind and sun exposure can lead to substantial differences in the yield damage caused by unfavorable climatic, pets and disease events, as can a few days difference in planting dates or the crop varieties grown.

Second, the scheme is subject to severe covariability problems. When the average yield is below the critical value in a region, all the insured farmers have to be compensated simultaneously. An unsubsidized insurer could hope to survive only if the scheme were to span a large number of regions with negative or positive but weakly correlated yields. The alternative of seeking commercial reinsurance seems unlikely given private insurers' reluctance to insure yields against a wide array of perils.

India introduced a national area-yield crop insurance scheme in 1985 (the Comprehensive Crop Insurance Scheme, CCIS), following a pilot phase from 1979 to 1984. The insurance is only available to farmers who borrow credit from financial institutions; and the indemnities are paid directly to the lending institution. Mishra (1994) provides evidence from Gujarat that the insurance did increase lending to small farmers, though repayment rates varied slightly. Unfortunately, the premium rates charged farmers are only a fraction of the rates needed to cover the indemnities paid, and even the small premiums charged (1-2% of coverage) are heavily subsidized by government. As a result, the financial performance of CCIS to date has been disastrous. The CCIS collected Rs 11,961 lakhs ($40 million) of premium during 1985-92, but paid Rs 90,163 lakhs ($303 million) indemnities over the same period (Joshi, nd). Considering that government subsidized about half the premium collected, CCIS paid Rs 15 for each rupee of premium collected from farmers. The insurance programme will clearly need to be reformed if it is to become a cost-effective risk management policy. Another variant of area-based insurance is regional rainfall insurance. In this case the insurance pays out whenever the average rainfall for a region falls below some critical value. Rainfall is easier to measure than average yield, and with modern satellite imagery it is now possible to assess the soil moisture content for a region with the minimum of field inspection. This information could be used to confirm rainfall readings, or even used directly as the insured peril.

Unlike regional yield insurance, rainfall insurance is not tied to the performance of specific crops, Since most farm families rarely depend on a single crop for their total income, the attractiveness of rainfall insurance should depend more on how the total insurance is limited to the occurrence of severe drought or floods, it is likely that in semi-arid or floodprone areas the indemnities would coincide with catastrophic income outcomes for many rural families.

Drought insurance could be marketed rather like lottery tickets, employing lowincome people to sell tickets on a commission basis. Unlike standard lotteries, however, all ticket holders would win a prize in a disaster year, bot no prize would be given in non-disaster years. There is no need to restrict the insurance to farm families, and all types of rural households might find it attractive. This is because a decline in farm incomes in drought or flood years usually leads to a sharp contraction in the rural non-farm economy, and the incomes of many workers and businessmen decline in tandem. Drought insurance faces the same covariability problem as regional yield insurance, but because it is limited to a specific weather peril, it might be much easier to obtain international reinsurance. This prospect could be enhanced if the scheme were run by a commercial bank or insurer within the country.

SPECIAL PROGRAMS FOR POVERTY ALLEVIATION

The relatively favorable performance of better endowed rainfed regions. Continued increases in productivity in such regions can be expected to help reduce poverty, particularly in combination with efforts to improve infrastructure and institutions critical for delivery of services. Likewise, widespread poverty is likely to remain for a long time in poorly endowed areas less likely to enjoy high agricultural growth and rising wages. For these areas even more than others, migration and development of the nonagricultural sector must provide the solutions. Singh (1990), von Braun (1995) and others, however, comment that while economic growth and diversification may be the solution to long term poverty alleviation, the process takes too long to solve today's poverty problems. Growth on the scale required cannot happen overnight, and it takes time for it to have a significant impact on poverty.

This is the rationale for the introduction of numerous special antipoverty programmes introduced in India beginning in the 1970s. The programmes are intended to alleviate poverty in the short term while seeking to stimulate economic growth that will reduce poverty in the long term. Of the many central government schemes initiated in India, the largest are the Integrated Rural Development Program (IRDP), introduced in 1978-79, and the National Rural Employment Program (NREP), which began in 1980. The IRDP mainly provided subsidies and credit to purchase livestock and milch animals, outlaying Rs 10 billion and covering about 3 million families per year in the mid 1980s. This programme, however, suffered from difficulties in targeting benefits to the poor and encouraging repayment of loans. Despite its large coverage, it paled in comparison with the needs of the estimated 260 million rural people.

The NREP focused mainly on employment, generating 350 million additional days of employment per year through Food for Work programmes that executed a wide variety of public works projects, including roads, drinking water projects, small scale irrigation works, soil and water conservation and afforestation.

Singh and Dantwala argue that employment programmes have a greater capacity to help the poorest people because they can be self-targeted, with less spillover than other measures such as food aid, and they can also create durable infrastructure that leads to development. Von Braun (1995) presents a conceptual model and detailed evidence of the positive contribution that employment programmes can make to poverty alleviation.

He shows why employment generation can be a superior alternative to other measures such as food subsidies, which benefit poor people but are expensive, distort food prices, and suffer significant leakage to nonpoor households due to targeting difficulties.

Employment programmes, on the other hand, can be self-targeted towards poor people whose only productive asset is their labor. Employment programmes can, in principle, be self-financed though taxation that pays not only for employment but productive assets and improved infrastructure that employment generates. In India, employment programmes have been used for constructing canal irrigation schemes, building roads and drinking water wells and buildings. Eventually some concern arose that the opportunities for productive, labor intensive infrastructure development were dwindling (Thomas Walker, pers comm.). The amount of unutilized canal irrigation potential dwindled, virtually every village had a school house and a health centre, and road work was confined mainly to maintenance, not construction of new roads.

With the increased focus on developing rainfed agriculture over the last decade, there has been an increased focus on using employment to support rainfed agriculture. Watershed development, including soil conservation bunds and ditches, check dams and tree planting, appeared to be a useful way to support the development of rainfed agriculture in distressed areas while also providing short term employment.

The focus on asset creation to support rainfed agriculture is attractive, but there is insufficient information regarding its effectiveness to date. Jackson (1992), in a worldwide study that included India, argued that the asset-creation impact of employment programmes was exaggerated due to the low quality of work they produce.

Kerr et al (1994) argue that in India, watershed development efforts linked to public employment programmes are unlikely to have a lasting impact on rainfed agriculture under current programme design. They present numerous field observations suggesting that soil and water conservation programmes based on employment programmes may create illusory gains.

This is because in order to obtain employment at the minimum wage, which in some cases exceeds the market wage, participants are encouraged to install mechanical structures that they may dismantle immediately afterward, or to plant trees that they will not protect. For this reason, the link between public employment and rainfed agricultural development may be more complex than first thought. In fact, in several countries, particularly in Africa and the Caribbean, employment subsidies have been separated from soil and water conservation programmes because the objectives of employment and natural resource management appeared to be in conflict.

The point is that long run asset-generation effects of employment programmes should not be taken for granted, particularly as related to rainfed agriculture. The papers in Von Braun (1995), for example, examine in great detail the benefits of employment but largely assume the durability of assets created. They devote more attention to whether such assets will result in equitable income distribution than the amount of such assets they will actually generate. Further research and field experimentation are needed to understand the nature of incentives they create and the steps needed to ensure that they are consistent with long term asset creation as well as short term poverty alleviation. Kerr et al (1994) suggest some ideas tried on the ground in India and Africa to achieve that objective.

Still, it may prove to be that some kinds of assets, like roads, wells and buildings can be created more effectively through employment programmes than others, like trees and soil conservation bunds. Other problems also remain with employment programmes. Jackson (1992), for example, argues that such programmes often are associated with high leakage. Unpublished ethnographic data from ICRISAT (1993, 1995) also cites villagers' complaints of leakages and favoritism in distributing employment benefits, but there is no evidence on the scale of the problem.

On the other hand, such government programmes will always be associated with some leakage, but that does not necessarily mean that they should be eliminated. In any case, there is room for improvement in making employment programmes more cost-effective. Finally, one important aspect of employment generation programmes is their capacity to be self-targeting, with minimal distortion to the rural economy. These objectives can be achieved if wages are kept slightly below the going market wage, and if they are concentrated in the slack season, when employment demand is low. If wages are low, only those people who really need them will participate.

If they are high — above the market wage, in particular — then employment will be rationed and some people who need it will not be able to get it. Ravallion, Datt and Chaudhuri (1990), for example, found that raising wages may cause a greater proportion of the poorest people who seek work to be excluded. This is so because as programme benefits become more attractive and have to be rationed, less powerful people will have a more difficult time bidding for them.

Low wages also ensure that small farmers who rely on hired labor during peak periods do not suffer from having to pay high wages due to labor scarcity created artificially through public employment programmes. Legislation in India requiring public employment programmes to pay the national minimum daily wage of Rs 22, though well-intentioned, may cause employment programmes to be less effective in reaching the poorest people, make them financially unaffordable, and damage the interests of small farmers by raising wages above the market wages that prevail in some areas.

GRASSROOTS ANTIPOVERTY INITIATIVES THAT SUPPORT RAINFED AGRICULTURE

Over the years, Indian NGOs have experimented with a wide variety of grassroots approaches to rural development. Some of these, honed by a relatively small number of high quality NGOs, appear to have promise for poverty alleviation and increased productivity even in marginal areas. The wide variety of approaches to community development defy easy categorization, but to be brief, we outline a stylized example of the experience to date.

Many NGOs find that informal savings cooperative groups are high on the list of villagers' priorities for development assistance. Savings groups are established loosely, with members selecting the composition of their own groups, and many groups, typically of homogeneous membership, often operate simultaneously in the same village.

Experience suggests that participation in these groups helps people generate capital, develop organizational skills, build villagers' confidence to work collectively, to seriously consider new investment opportunities, and to act where they never bothered to previously (James Mascarenhas, OUTREACH, personal communication; Parthasarathy, 1994). Active local groups can stimulate psychological incentives that previously were stifled by cultural or political constraints.

Mascarenhas, et al (1991) and Fernandez (1993) discuss links between promotion of such groups and the development of rainfed agriculture through watershed management.

First, participation in credit groups can help villagers save funds to invest in rainfed agriculture. One common experience in MYRADA's and Oxfam's thrift group projects is that even without any financial assistance, within a year or two participants find that their greatest challenge is to figure out how to spend their savings.

Second, developing and strengthening cooperative groups prior to tackling watershed development helps villagers build organizational and conflict resolution skills that are important in watershed management.

Third, the groups can serve as a focal point for spreading awareness about the benefits of watershed development. In MYRADA watershed programmes, existing cooperative groups developed to solve an unrelated set of problems become a focal point for development of rainfed agriculture.

Small group credit generation programmes have multiplied rapidly in south India. MYRADA's and Oxfam's thrift projects, for example, have tens of thousands of participants, and they are spreading rapidly. Information about these projects is limited, and their potential to promote rainfed agriculture through indirect means is not well understood. They are worthy of further attention, both to understand the role they can play in alleviating poverty and promoting rainfed agriculture, and to understand how to extend such an approach to a state or national scale.

POVERTY-ENVIRONMENT INTERACTIONS IN AGRICULTURE

Agriculture accounts for most land use in the developing countries, and as such is probably the single most powerful influence on environmental quality. Characteristics of agricultural production systems, such as location in the landscape, type and timing of cultivation, agrochemical input use, field burning and fallow use, and livestock practices, affect many environmental variables, particularly water quality and flow, soil quality and movement, natural vegetative cover and biodiversity, at local, regional and international scales. At the same time, agriculture remains the principal livelihood of poor people in developing countries, and particularly the rural poor, and is the major "engine" of economic growth in a majority of developing countries. Projections of rural population growth, agricultural expansion and intensification, and poverty in the next few decades suggest a potentially serious conflict between environmental quality and poverty reduction in agricultural regions. Indeed, assumptions about the relationships between poverty, agriculture and environmental degradation have driven major policy initiatives in both the agriculture and environment sectors during the last century. Yet these assumptions are often not empirically supported and there is continued debate among academics, development practitioners, and advocates for the poor and for the environment, about directions and nature of causality and appropriate policy response.

The objective of this paper is to highlight key elements of this debate, assess current evidence and experience, and draw from this implications for design of agriculture- and related policies which can reduce the impacts of resource degradation on poverty, reduce the impacts of poverty on resource degradation, and contribute jointly to poverty reduction and environmental rehabilitation. The following section defines poverty and environment issues related to agriculture. The third section describes different aspects of their relationship, and presents a framework for assessing change processes and policy action. The fourth section discusses eight factors which explain why, under some circumstances, negative poverty-environment linkages occur and in others they are avoided or positive adaptations are made. Based on this evidence, the fifth section suggests a number of policy strategies which could jointly address poverty and environment objectives in the agriculture sector. The final section discusses remaining gaps in our knowledge about the poverty-agriculture-environment nexus and general challenges that deserve particular attention as the new century begins.

AGRICULTURE, THE POOR AND ENVIRONMENTAL DEGRADATION

Discussion in this paper is confined to environmental issues associated with agriculture. Concerns relate mainly to the sustainability of the resource base for agricultural production (*e.g.*, soil quality, access to irrigation water), the protection of biodiversity and related habitats, and environmental services of resources influenced significantly by agricultural land use (*e.g.*, watershed function, carbon sequestration).

These concerns are not trivial. Degradation of soil and vegetative resources already threatens agricultural productivity, biodiversity and water quality and availability in many "hot spots" in the developing world. Soils in approximately 16 percent of developing country land area, and a higher proportion of cropland and drylands, have degraded moderately or severely since mid-century, mainly through soil erosion, nutrient depletion and salinization. At least 28 countries with a total population exceeding 300 million people face water stress today, and demand is growing rapidly even as water contamination due to agriculture and rural domestic uses increases. Declines in agrobiodiversity increase disease and pest problems. Agricultural expansion and intensification is the leading cause of species loss and depletion of natural vegetation, including bamboos, palms, and grasses harvested for human and domestic animal use. The poor are implicated in only a part of this degradation and its consequences. Wealthier farmers, agricultural investors, and multinational corporations typically control much more total land area than the poor, and have played a prominent role in large-scale clearing of natural vegetation, over-use of agro-chemicals, large-scale degradation of grazing lands, over-exploitation of soils for export production. However, the poor do play an important role in unsustainable agricultural intensification, expansion of farming into marginal lands, and over-exploitation of vegetation. The consequences of degradation for the poor also tend to be more serious, because they lack assets to cushion the effects.

"Poverty" is a concept which indicates absolute or relative welfare deprivation. It may be defined in terms of private consumption alone, or extended to include access to common property resources and state-provided commodities, personal assets or wealth, and even subjective assessments of "dignity" and "autonomy". IFAD identified five types of rural poverty which have different links with agriculture and the environment. Material deprivation and alienation cause "interstitial poverty", or pockets of poverty surrounded by power and affluence. "Peripheral poverty" is similar, but found in isolated, marginal areas. "Overcrowding poverty" is material deprivation arising from population pressure and limits on resources. Vulnerability to natural calamities, labor displacement and insecurity produces "traumatic or sporadic poverty", which can be transitory. Isolation, alienation, technological deprivation, dependence and lack of assets are signs of "endemic poverty."

Reardon and Vosti's typology of poverty is explicitly linked to environment. They examine the asset portfolio of the rural poor, in terms of (1) natural resources, such as water, ground cover, biodiversity of wild and domestic fauna and flora, and soil; (2) human resources, such as education, health, nutrition, skills, number of people; (3) on-farm resources, such as livestock, farmland, pastures, reservoirs, buildings, equipment, financial resource); (4) off-farm resources, including local off-farm physical and financial capital; (5) community-owned resources such as roads, dams and commons; and (6) social and political capital.

Where assets are not completely fungible (markets are absent, underdeveloped or constrained), asset-specific poverty can influence livelihood activities and investment decisions. They note that "welfare poverty" criteria can miss a potentially large group of households that are not "absolutely poor" by the usual consumption-oriented definition, but are too poor—in that their surplus above the minimum diet line is still too small to make key conservation or intensification investments necessary to prevent their land use practices from damaging the resource base or leading them to push onto fragile lands. They prefer use of a measure of "conservation-investment poverty", the cut-off for which is site specific, a function of local labor and nonlabor input costs and the types of investment that are needed for the particular environmental problems or risks faced.

Data Limitations

Debates about poverty-environment linkages have been difficult to resolve in part due to data limitations. There have been few longitudinal studies linking poverty and resource quality in agricultural system. Most of those which exist have been reconstructed from oral history and from archival, remote sensing or time series survey data originally collected for other purposes, which do not examine poverty and agricultural production systems in a geographically explicit context. Emphasis has been more on measuring poverty, than explaining why people are poor and the relative importance of environmental conditions or environmental degradation. Most assessments have used aggregate, macro-scale data, when the key questions in fact require micro-level analysis.

Index